Biotechnology Explorations:
Applying the Fundamentals

Biotechnology Explorations:
Applying the Fundamentals

Judith A. Scheppler
Grainger Center for Imagination and Inquiry
Illinois Mathematics and Science Academy
Aurora, Illinois

Patricia E. Cassin
Biology Department
Nassau Community College
Garden City, New York

Rosa M. Gambier
Biotechnology Learning Center
Suffolk County Community College
Selden, New York

Washington, D.C.

American Society for Microbiology
1752 N Street NW
Washington, DC 20036

Library of Congress Cataloging-in-Publication Data

Scheppler, Judith A.
Biotechnology explorations: applying the fundamentals / Judith A. Scheppler, Patricia E. Cassin, Rosa M. Gambier.
p. cm.
ISBN 1-55581-178-7 (softcover)
1. Biotechnology—Study and teaching. 2. Biotechnology—Laboratory manuals. I. Cassin, Patricia E. II. Gambier, Rosa M. III. Title.
TP248.22 .S347 2000
660.6′078—dc21

99-089447

10 9 8 7 6 5 4 3 2 1

Printed in the United States of America

Cover and interior design: Susan Brown Schmidler

Contents

Preface

Biotechnology is the use or manipulation of any living organism with the intended purpose of improving something. In its broadest sense it is as old as we are. Before realizing what they were doing, humans were using biotechnology by culling breeding stocks for preferred characteristics, collecting seeds and propagating plants to enhance food production, and using microorganisms for the production of bread, beer, and cheese. Biotechnology is now a recognized field, in which scientists, using advanced techniques, directly manipulate genes. Information resulting from recombinant DNA technology and cloning, and the Human Genome Project, plus the ability to intervene and alter specific genes has presented the scientific and nonscientific community with the challenge of managing, understanding, and ethically using this growing mass of knowledge.

Teaching the Basics with Opportunities for Inquiry

In producing this manual, we recognize the importance of including biotechnology in the course work offered to all undergraduates. Our goal was to develop and select a collection of activities that would pique the students' interest and provide them with opportunities for independent study. Investigations are enhanced with sidebar comments detailing information to help ensure success. Students are guided with formulation tables to prepare some of the materials for their own laboratory exercises. In addition, each investigation includes suggested readings, laboratory language, and analysis questions that help direct student thinking. Inquiry-based activities are chosen for their safety, reliability, and flexibility. We have chosen a five-kingdom approach although we are fully aware of the ongoing classification controversies challenging the adoption of kingdoms in favor of domains. Students will be able to use some of the application exercises to gain insight into the role of biotechnology in resolving classification questions while giving them a chance to compare genome sizes and homologies among representative organisms. Other helpful information is found in the sidebar comments to the protocols.

Organization—Fundamentals and Applications

Biotechnology Explorations: Applying the Fundamentals offers a broad range of activities that are easily adapted to fit the time frame of a one- or two-semester laboratory course. The exercises are organized into the fundamental laboratory activities and applications of these protocols. "The Fundamentals" exercises are designed to reacquaint students with the characteristics of the nucleic acids, review the basics of solution preparation, and teach the techniques basic to the current work in biotechnology. Transformation, restriction enzyme analysis, and polymerase chain reaction protocols are included here. "Applying the Fundamentals" offers students a choice of applications and extensions of the fundamental techniques. Students investigate gene structure, plasmid library construction, protein expression, and DNA sequencing, using more complicated protocols such as ligation and Southern and Western blots.

We encourage instructors to personalize the exercises here, to pick and choose the particular type and level of exercise needed to teach the basics to beginning students, or to select the exercises from "Applying the Fundamentals" to challenge the students who have a strong grasp of basic biotechnology concepts and protocols. The flexible design of the material is intended to give instructors a source of stand-alone biotechnology experiments that meet individual student needs.

A Collaborative Effort

Biotechnology Explorations: Applying the Fundamentals is the result of a unique collaboration between research and education. The authors represent a 4-year baccalaureate program, a 2-year community college program, and secondary education. Although we each have our own specific teaching environment and come from very different academic backgrounds, we share the belief that biotechnology instruction is central to science education and that how it is taught to students matters as much as the content itself. We began the project because of a search for biotechnology protocols that would allow us to reach our diverse student populations. Most of the biotechnology protocols available were too involved or required extra laboratory time and materials that were not easily available. In addition, most protocol collections used the same level of organism throughout. We wanted students to be able to investigate the similarities and differences among the kingdoms. Finally, we needed a source of straightforward, reliable protocols and extension activities that were safe and affordable. We found a workable solution by combining our research and teaching experience. Working at three different academic levels gave us the chance to test each activity in a variety of teaching situations. Our collaboration has been exceptionally rewarding. We hope that instructors and students enjoy the manual and that it is just the beginning for them of a continued involvement in one of the most powerful areas of science.

Acknowledgments

We thank our families, colleagues, and friends for their support, especially the Department of Biochemistry, SUNY at Stony Brook, and Stephen Beck and Catherine D. Kelly, Biology Department, Nassau Community College. We thank Robert Stone, Brad Boyer, Jennifer Black, James O'Connor, and Paul Dinerstein from the Biology Department at Suffolk County Community College. We are indebted to Don Dosch and Sue Styer at the Illinois Mathematics and Science Academy. In addition, we thank Pat Bany from Jericho Public Schools.

Special thanks go to Karen Trevors, Karen Pike, and Barbara Moffatt at the University of Waterloo for reviewing our work in a course and classroom setting.

We gratefully acknowledge the support provided by LIGASE, The Long Island Group About Science Education, a grant from the National Science Foundation (DUE #9602450) under which we began our collaboration and some of this work.

The following people and companies generously provided reagents and technical assistance: Kat Davis and Athena Kobal at Promega, Ron Mardigian at Bio-Rad, Sherry Levitt at New England BioLabs, Fermentas, Fisher Scientific, and Cedarburg Laboratories.

Most of all, we thank our students, both high school and college, and the high school and community college faculty whom we have taught, for their invaluable comments and patience as we developed these laboratory exercises.

Finally, we are indebted to ASM Press, including Jeff Holtmeier, Susan Birch, Ken April, Mary McKenney, and Pamela Lacey, for their various contributions to this project.

The Double Helix is indeed a remarkable molecule. Modern man is perhaps 50,000 years old, civilization has existed for scarcely 10,000 years, and the United States for only just over 200 years; but DNA and RNA have been around for at least several billion years. All that time the double helix has been there, and active, and yet we were the first creatures on earth to become aware of its existence.

Francis Crick
What Mad Pursuit

Laboratory Safety

Teaching and learning in the biotechnology laboratory may involve some degree of hazard under the very best of circumstances. Recently designed equipment, newly developed reagents, and biological materials are an intrinsic part of nucleic acid and protein laboratory exercises. Students are learning novel techniques, using reagents and biological materials that may pose some degree of risk, and operating new equipment in the confines of the classroom setting. Although one cannot perform hands-on laboratory exercises without using equipment, reagents, and biological materials, students and instructors are entitled to work in the safest, most pleasant laboratory classroom environment possible.

"Situational awareness," a phrase borrowed from professional airline pilots, offers a guiding theme for maintaining laboratory safety in the face of inherent risks. As the airplane pilot must be aware of his or her entire surroundings with constant vigilance directed toward safety of self and others, so must biotechnology laboratory students and instructors do the same. The pilot must always know "where he is, where he is going, and how to get there." Translated to the laboratory setting, students must be focused on the task at hand; must understand the equipment, reagents, and biological materials in use; and must be given and follow clear directions to conduct an experiment. At the same time, every student must be aware of his or her lab partners and the protocols they are performing. Instructors must set an example by knowing and cheerfully and enthusiastically promoting safety regulations. Most important, instructors must be attentive to their students and ready to provide guidance.

Most institutions have detailed safety guidelines for working with and disposing of chemical and biological reagents and with living microorganisms. Consult your Environmental Health and Safety Officer for specific requirements in effect at your institution. Instructors and students are obliged to know and follow the rules promulgated by their institution. Many institutions require a laboratory safety training session at the beginning of each laboratory science course. A student safety quiz and student safety contract may be provided by the institution or by each individual instructor as part of lab safety training. (Samples are given below.)

With all of the above in mind, safety issues may be divided into four areas:

- general dress and deportment
- physical hazards
- chemical hazards
- biological hazards

General Dress and Deportment

Report all accidents to the instructor!

- Keep personal belongings and other materials at the lab bench to a minimum. Most personal belongings should be placed in a common coat rack or storage area of the classroom away from the lab benches.

- Laboratory coats or other suitable protective clothing must be worn at all times to avoid soiling or contaminating one's street clothing.
- Hair must be pulled back and secured away from the face to minimize possible ignition when working with the Bunsen burner and to avoid chemical or microbial contamination of one's person.
- Hats may not be worn: they may pose a fire hazard.
- Closed-toe shoes are required; sandals or similar shoes are not permitted since they will not protect against chemical or microbial spills.
- Chemical splash goggles are required as demanded by specific protocols.
- Ultraviolet (UV) protective goggles are required as demanded by specific protocols.
- Disposable exam gloves are required when handling microorganisms, enzymes, nucleic acids, and other biochemical reagents.
- Jewelry must be kept to a minimum, especially rings and bracelets that could tear gloves or come in contact with biological or chemical reagents.
- Food, beverages, chewing gum, cosmetics, hand lotions, and creams are **never** permitted in the laboratory: reagents or microbes could accidentally be introduced into the mouth or onto the skin.
- Students must thoroughly understand experimental procedures before coming to the laboratory class: proper planning prevents poor performance.
- Students must question the instructor when directions appear unclear, or when they are in doubt about how to proceed.
- Student behavior that could introduce risk of injury or laboratory accident is grounds for dismissal from the classroom.
- Each person must wash his or her hands before leaving the laboratory.

Physical Hazards

In biotechnology laboratories, the major physical hazards are posed by fire, autoclaved and hot solutions, broken glassware and sharp instruments, electrical equipment and power supplies, and UV radiation.

Fire

- Know the locations of the fire blanket, safety shower, and fire extinguishers in the laboratory.
- Know the exits from the laboratory.
- Know the location of the master shutoff controls for gas and electricity.
- Use a heat block, hot plate, or water bath for heating liquids.
- Heating biological samples to high temperatures poses the risk of burns. Use tongs, forceps, or heatproof gloves.
- Superheated agarose solutions can boil over, causing burns. Care must be taken when removing them from the hot plate or microwave oven. Use tongs and/or heatproof gloves.
- In case of fire to clothing or person, immediately "stop, drop, and roll." Alternatively, use the safety shower or fire blanket to extinguish flames.
- Use a fire extinguisher to extinguish burning property or lab work, **never people.**

Bunsen burner

Bunsen burners have the potential for causing severe burns and starting fires if used improperly.

- Know how to light and use a Bunsen burner properly.
- Never leave a flaming Bunsen burner unattended.
- Flame instruments and tubes carefully to reduce aerosol formation.
- Never work with a Bunsen burner on or very near flammable material.

Autoclave
Recently autoclaved solutions, such as sterilized microbial media, are very hot when removed from the autoclave and could cause potentially severe burns.

- Use heatproof gloves to remove media from the autoclave.
- Similarly, when a recently used autoclave door is opened, steam may escape, causing burns. Care must be taken when working with the autoclave to minimize these risks.

Broken glassware
Lacerations from broken glass are of great concern, especially if the glassware contained live microbial cultures. Decontamination is discussed below under "Biological Hazards."

- Broken glass must be disposed of according to your institution's directives, ordinarily into a special "sharps" container for later removal and disposal.
- Broken glass is **never** placed in the routine trash containers.
- Pick up broken glass carefully, using gloved hands to manipulate paper towels or dustpan and brush.
- Microbially contaminated broken glassware must be disinfected before cleanup.

Use of razor blade or other sharp instrument
Occasionally, it is necessary to use a razor blade to cut DNA fragments from a gel or to use sharp instruments to assist with tissue dissociation into single cells.

- Work carefully, wearing gloves that fit properly.
- Do not use excessive force.

Electrical equipment and gel electrophoresis apparatus
Electrical equipment and power supplies used in the biotechnology laboratory have the potential to cause electrical shock injuries and fire.

- Always check the electrical cords of each piece of equipment; do not use any equipment with frayed cords.
- Electrical equipment should not be used in or near water.
- Set all equipment to the "off" position before plugging it into the electrical outlet.
- In case of sparks or fire in electrical equipment, turn off the electricity at the master shutoff. Use a fire extinguisher when appropriate.
- Never plug or disconnect an electrophoresis apparatus into or from a power supply unless the power is off.
- Never remove the lid of an electrophoresis apparatus unless the power supply is turned off.

UV transilluminators
UV radiation can damage the skin and eyes.

- Use a UV transilluminator with a UV-blocking lid.
- Use a UV-blocking face mask or eye goggles if working with a transilluminator that is not equipped with a UV-blocking lid.

Chemical Hazards

- Know the location of the material safety data sheets (MSDS) for all chemical reagents used in the laboratory. All MSDS should be kept together in a binder or other easily displayed form in a readily accessible place in the laboratory. MSDS are the most complete source of information about physical and chemical properties, hazards, spill handling procedures, and first aid for each reagent.

- Instructors must discuss specific chemical hazards before beginning the lab procedure in which these chemicals are used. **It is the instructor's responsibility to inform students about chemical safety.**
- Students are responsible to know and understand the hazards involved and to ask for clarification if they do not understand. **It is the student's responsibility to learn and implement safe procedures for handling chemicals.**
- **Never pipette anything by mouth.** Always use a mechanical pipette aid device.
- Know the location and use of the chemical fume hood. Use the fume hood when working with highly volatile reagents.
- Generally, chemical wastes of any kind may not be disposed into sink drains.
- Know the location and use of chemical waste disposal containers according to your institution's directives.
- Know the location of all sinks, the eyewash station, and the safety shower in the laboratory.
- Should chemicals be spilled on the clothes or skin, remove contaminated clothing and wash affected skin for 15 minutes in a sink or safety shower.
- If chemicals are splashed into the eyes, use the eyewash station to wash eyes, eyelids, and face for 15 minutes. Hold the eyelids open.
- Never remove reagents from the laboratory.

Some of the specific chemicals used in biotechnology laboratories that pose hazards are listed below. The information included here is not intended to substitute for a thorough reading of the appropriate MSDS, which is the responsibility of everyone working with any chemical materials.

The use of premixed solutions can greatly minimize exposure to powders (e.g., acrylamide, sodium dodecyl sulfate, ethidium bromide) and can reduce time needed for prelab preparation (e.g., nitroblue tetrazolium, 5-bromo-4-chloro-3-indolylphosphate). One must carefully read all MSDS sheets that accompany premixed reagents; **some solvents used to dissolve chemicals in premixed solutions also pose their own specific hazards.**

- **Acridine orange:** Possibly mutagenic. Disposable exam gloves, lab coats, and chemical splash goggles are required. Do not work with solid acridine orange crystals; purchase premixed solutions.
- **Acrylamide and bisacrylamide solutions:** Mutagenic, carcinogenic, and neurotoxic. Disposable exam gloves, lab coats, and chemical splash goggles are required. Do not work with solid acrylamide crystals; purchase premixed solutions.
- **Ammonium persulfate:** Strong oxidizer; keep away from heat and combustible materials. Can cause severe respiratory, digestive, and skin irritation, with possible burns. Disposable exam gloves, lab coats, and chemical splash goggles are required.
- **5-Bromo-4-chloro-3-indolylphosphate (BCIP):** Possible carcinogen. Must be used in a fume hood. Chemical-resistant gloves, lab coats, and chemical splash goggles are required. Minimize hazards by purchasing premixed solutions.
- **5-Bromo-4-chloro-3-indolyl-β-D-galactopyranoside (X-Gal):** Dissolved in DMF.
- **Chloroform:** Causes eye, skin, and mucous membrane irritation. Suspected carcinogen. Prolonged exposure may cause damage to heart, liver, kidneys, and nervous system. Must be used in a fume hood, especially large volumes. Chemical-resistant gloves, lab coats, and chemical splash goggles are required.
- ***N,N′*-Dimethylformamide (DMF):** Carcinogenic; toxic to kidneys and liver. Must be used in a fume hood. Chemical-resistant gloves, lab coats, and chemical splash goggles are required.

- **Dithiothreitol (DTT):** Causes eye and skin irritation; irritating to mucous membranes and upper respiratory tract. Use in a well-ventilated room. Disposable exam gloves, lab coats, and chemical splash goggles are required.
- **Dimethyl sulfoxide (DMSO):** Combustible liquid and vapor; avoid heat. Irritating to skin, eyes, and respiratory tract. Chronic exposure may affect blood, skin, kidneys, and liver. Use in a well-ventilated room. Chemical-resistant gloves, lab coats, and chemical splash goggles are required.
- **Ethanol (ethyl alcohol):** Flammable liquid; avoid contact with strong oxidizers, heat, and strong alkalis and acids. Prolonged exposure can cause central nervous system depression. Disposable exam gloves, lab coats, and chemical splash goggles are required.
- **Ethidium bromide solutions:** Ethidium bromide is considered a mutagen. Disposable exam gloves, lab coats, and chemical splash goggles are required. Minimize hazards by purchasing premixed solutions.
- **Formic acid:** Corrosive to skin, eyes, and mucous membranes. Prolonged exposure may cause liver and kidney damage. Work in a fume hood. Chemical-resistant gloves, lab coats, and chemical splash goggles are required.
- **Guanidine hydrochloride:** Irritating to skin, eyes, and respiratory tract. Harmful if swallowed; can cause gastrointestinal and neurological disturbances. Disposable exam gloves, lab coats, and chemical splash goggles are required.
- **Hydrochloric acid (HCl):** Highly corrosive and causes severe burns on eye and skin contact or on inhalation of gas. Chemical-resistant gloves, lab coats, and chemical splash goggles are required. Minimize hazards by working with diluted solutions instead of concentrated acid.
- Isopropanol: Irritating to eyes, skin, mucous membranes, and upper respiratory tract. Prolonged exposure can cause central nervous system depression. Disposable exam gloves, lab coats, and chemical splash goggles are required.
- **Methanol (methyl alcohol):** Flammable liquid; avoid contact with strong oxidizers, heat, and strong alkalis and acids. Prolonged exposure can cause central nervous system depression. Disposable exam gloves, lab coats, and chemical splash goggles are required.
- **Ninhydrin:** Irritating to eyes and skin. Work in a fume hood. Chemical-resistant gloves, lab coats, and chemical splash goggles are required.
- **Nitroblue tetrazolium (NBT):** Irritating to mucous membranes and skin. Use in a well-ventilated room. Disposable exam gloves, lab coats, and chemical splash goggles are required.
- **Sarcosine (*N*-lauroylsarcosine sodium):** May be harmful by inhalation, skin absorption, or ingestion. May be irritating. Disposable exam gloves, lab coats, and chemical splash goggles are required.
- **Sodium dodecyl sulfate (SDS):** Irritating to eyes and skin. Harmful by contact with skin or ingestion. Disposable exam gloves, lab coats, and chemical splash goggles are required.
- **Sodium hydroxide (NaOH):** Highly corrosive; causes severe burns to skin, eyes, and mucous membranes. Chemical-resistant gloves, lab coats, and chemical splash goggles are required. Minimize hazards by working with diluted solutions instead of concentrated base.
- **Sulfuric acid:** Highly corrosive and causes severe burns on eye and skin contact or on inhalation of gas. Chemical-resistant gloves, lab coats, and chemical splash goggles are required. Minimize hazards by working with diluted solutions instead of concentrated acid.
- ***N,N,N′,N′*-Tetramethylethylenediamine (TEMED):** Flammable; harmful liquid and fumes. Can cause severe irritation and burns. TEMED is extremely destructive to tissue of mucous membranes, upper respiratory system, and skin. Must be used in a fume hood. Chemical-resistant gloves, lab coats, and chemical splash goggles are required.

Disposable exam gloves, lab coats, and goggles must always be worn when working with these compounds.

Biological Hazards

Most biological hazards arise from working with cultures of microorganisms. Review the general guidelines for dress and deportment above.

- Disposable latex or vinyl exam gloves, lab coats, and chemical-biological safety goggles must be worn at all times when working with microbial cultures.
- Disinfect your work area at the beginning of the lab exercise and again at the end of the lab exercise with 10% chlorine bleach solution or other appropriate disinfectant.
- Avoid all finger-to-mouth and finger-to-eye contact. Do not put pens or pencils into your mouth.
- Flame instruments and tubes carefully to reduce aerosol formation.
- Turn off the Bunsen burner when it is not in use.
- Never remove reagents, media, or cultures from the laboratory.
- Students should handle only their own cheek cell samples to avoid the possible transmission of infectious agents. It is unlikely that many infectious agents will remain viable after the detergent steps, but a few bacteria are resistant to soaps.
- Wash your hands thoroughly with antimicrobial soap before leaving the lab.

Besides working carefully with bacteria, one must dispose of bacterially contaminated waste materials properly. Although most plasmids and bacterial vectors used in the teaching laboratory have been genetically altered to pose little to no risk of transmission or infection if released into the environment, prudence dictates that **all plasmids, DNA solutions, and vectors be treated as contaminated waste.** The best method for discarding bacterial waste is to autoclave all materials before disposal. **Anything** contaminated with bacteria must be autoclaved or disinfected. **It is important to follow the procedures for handling contaminated waste mandated at your institution.** General guidelines follow:

- Discard cultures and contaminated waste **only** in the appropriate biohazard container indicated by your instructor. Special bags for collecting material to be autoclaved are frequently used in the lab.
- **Do not** put uncontaminated materials in the containers for contaminated waste.
- Immediately report all accidents involving cultures to your instructor.
- Culture spills must be decontaminated with 10% bleach or other suitable disinfectant.
- **Nothing** that has been in contact with bacteria goes into the regular trash, the laboratory trash, or the sink.

Accidents involving cultures must be reported to the instructor immediately.

- If a culture spills onto your work area or belongings, cover the contaminated materials with a single layer of clean paper towels, then soak with either 10% chlorine bleach solution or other suitable disinfectant. Allow the spill to stand for 10 to 15 minutes, then pick it up with paper towels and dispose of it in the contaminated-waste container.
- If a culture spills onto your skin, use an appropriate antiseptic, such as povidone-iodine, to clean your skin. Then wash with an antimicrobial soap.
- If broken glass is contaminated with microbial culture material, cover with a single layer of clean paper towels, then soak with either 10% chlorine bleach solution or other suitable disinfectant. Allow the glass to stand for 10 to 15 minutes, then carefully pick it up with paper towels, or dustpan and brush if appropriate, and dispose of it in the sharps container.

References and General Reading

Anonymous. 1987. Centers for Disease Control guidelines. *Morbid. Mortal. Weekly Rep.* **36**(Suppl. 25)**:**55–105.

Cassin, K. 1999. Personal communication.

Trammell, G., J. E. Pingel, J. Fisher, C. Josefson, M. Jackson, S. Vaughn, S. Matthews, and E. Ziebarth. 1995. *Guidebook for Science Safety in Illinois: a Safety Manual for Illinois Elementary and Secondary Schools.* Illinois State Board of Education Center for Educational Innovation and Reform, Division of Intermediate and Secondary Level Support, Springfield, Ill.

Internet Sites

http://hazard.com or http://siri.uvm.edu (an excellent site for MSDS information as well as other safety materials)

http://hazard.com/links.html (provides links to university safety programs posted on the World Wide Web)

SAMPLE SAFETY QUIZ

NAME ______________________________________ DATE ______________

1. Draw and label a floor plan of the laboratory, showing the instructor's desk, all the lab benches, your place in the lab, and the following:
   - exits
   - MSDS book
   - safety shower
   - eyewash station
   - sinks
   - fire alarm fire extinguisher
   - fire blanket
   - master gas shutoff
   - master electricity shutoff
   - fume hood
   - first-aid kit
   - sharps container
   - chemical waste container(s)
   - biologically contaminated-waste container
   - routine garbage container
2. What part of your body is most important to protect? How do you provide protection?
3. Why must food, beverages, chewing gum, cosmetics, and creams **not** be used in the lab?
4. Why does appropriate clothing include lab coat, gloves, closed-toe shoes, and no hats, and why must hair be restrained? Explain the rationale for each.
5. What is the correct way to handle each of the following situations?
   - **a.** A chemical is splashed into your eye.
   - **b.** A culture spills onto your lab coat.
   - **c.** You cut your hand on a broken culture tube.
   - **d.** Your notebook catches fire.

SAMPLE LABORATORY SAFETY CONTRACT

I, ______________________________, understand that much of the responsibility for laboratory safety is in my hands. I have read, understand, and agree to follow the safety regulations promulgated by my instructor. I understand that violation of these rules may result in expulsion from the laboratory.

Signed: ______________________________ Date: ______________

Student ID Number: ________________ Course: ____________________

Instructor: ______________________

The Fundamentals

SECTION I

Getting Started

EXERCISE 1

Documentation of Laboratory Experiments

Key Concepts

1. Laboratory protocols and experiments must be dated and recorded, in indelible waterproof ink, as they are performed.
2. A hard-bound laboratory notebook with consecutively numbered pages is required.

Goals

By the end of this exercise, you will be able to

- record protocols and data accurately and completely
- understand the importance of proper documentation

Introduction

Creativity may exist in the mind and soul, but documentation of experimental design must be preserved in a format that will withstand scrupulous inspection. Careful recording of protocols and results enables easier replication of experiments in the same lab or different labs. A table of contents in the front of the notebook simplifies finding information or cross-referencing several experiments in the same book or in different books. To refute any possible criticisms of a scientist's methodology, dated protocols, computations, experimental conditions, and results must be recorded in indelible ink, in a hard-bound notebook, as they are performed. If a researcher makes a mistake in computations or recording, he or she draws a single line through the inaccuracy and initials and dates the change. Erasures are never made, nor are errors obliterated. A previous "error" may turn out to be correct or may provide valuable information if an experiment does not yield the expected results.

Because it is difficult to alter data recorded in this manner, researchers can use a lab notebook to establish primacy of discovery or to apply for copyright and patent protection.

Many styles of laboratory notebooks are available. Size, page numbering style, and lines or grids on each page are options one might choose when selecting a suitable book. Several examples are pictured in Fig. 1.1 and 1.2.

Lab Language

computations
data
documentation
protocol
record
results

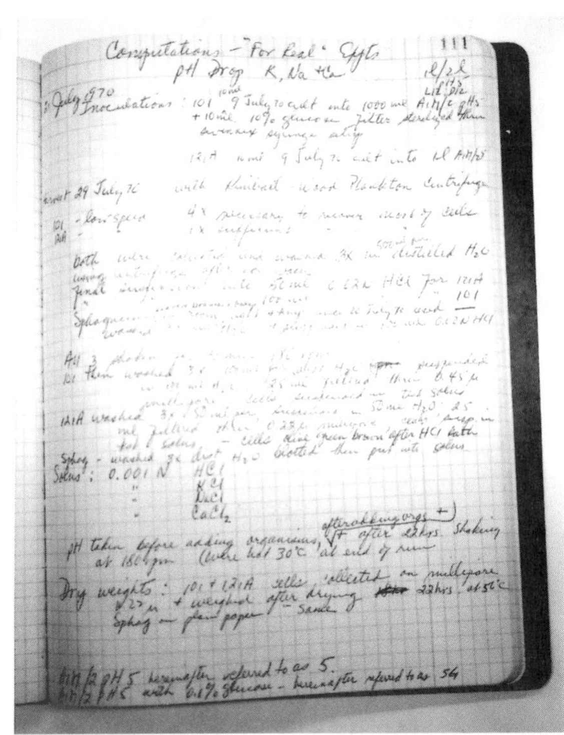

Figure 1.1 These are pages from an older computation (comp) book. Results are recorded in the beginning of the book with a reference to the calculations and preparations, which are recorded in the second half of the book. This method allows results to be easily read without obfuscation by complicated notes regarding the experimental setup.

Figure 1.2 These are pages from a newer comp book. Formulas for defined algal media are shown, with calculations for specific quantities written on consecutive pages. See the error in calculations noted at the bottom of page 067.

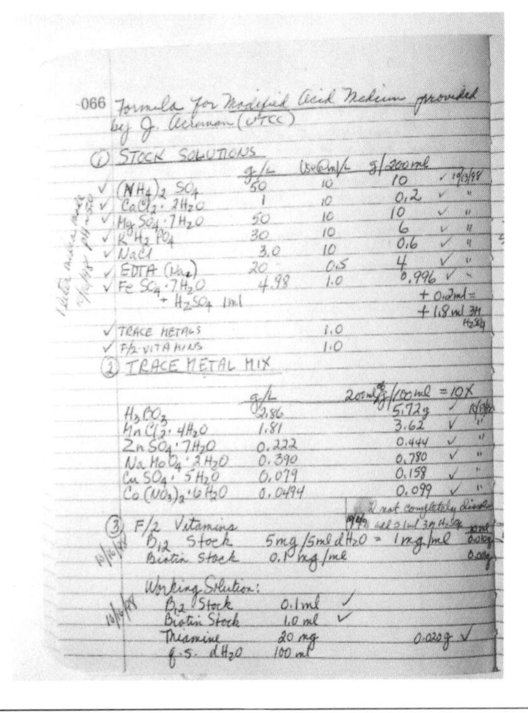

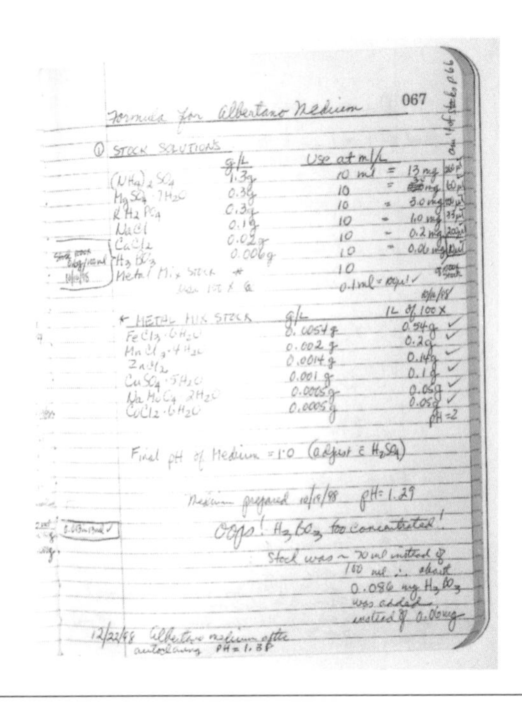

Materials

Materials per pair of students
indelible blue, blue-black, or black ink pens
laboratory notebook

Protocols

PROTOCOL 1 RECORDING COMPUTATIONS

1. Use an indelible ink pen.
2. Write today's date at the top of the numbered page you are using. Do not skip pages, and use lab book pages in numerical order.
3. Choose a solution or medium formulation you wish to prepare. Alternatively, you may elect any exercise or protocol to record.
4. Copy the formula, and write the calculations for the quantity you wish to prepare.
5. As you measure each component, check it off in your notebook.
6. Record any deviations you make from the formula or protocol.

Some researchers record everything in strict date order, starting with page 1. Others will use the second half of the notebook for computations, reserving the first half for recording results. Each results page references the computation page numbers.

PROTOCOL 2 RECORDING RESULTS

1. Choose any experiment. Exercise 6 is one suggestion.
2. Think about how you want to present your results. The format should be clear to any reader as well as to you.
3. Write today's date at the top of the numbered page you are using. Do not skip pages, and use lab book pages in numerical order.
4. Draw the appropriate chart or table as necessary.
5. Record your results as you obtain them. Include commentary where appropriate.

General Reading

Kevles, D. J. 1998. *The Baltimore Case: a Trial of Politics, Science, and Character.* W. W. Norton & Co., New York, N.Y.

Discussion

1. List the elements required for proper documentation of laboratory procedures.
2. Discuss the importance of authentic original documentation of experiments.

EXERCISE 2

Use of Equipment

Key Concepts

1. Equipment in the laboratory is expensive and must be used properly to ensure the best possible results in an experiment.
2. Listen carefully and follow the instructor's directions for use. Ask questions if you do not understand directions for use of equipment.
3. If something does not seem to be functioning correctly, if it makes an odd noise, or if you have never used a piece of equipment before, ask your instructor for directions.

Goals

By the end of this exercise, you will be able to

- locate equipment manuals in the laboratory
- use micropipettes correctly and accurately to measure small volumes
- correctly use a centrifuge and determine *g* force
- use a balance to measure desired quantities of materials
- determine and adjust the pH of a solution

Introduction

Pipettes

Pipettes are used to measure the volumes of liquids. Glass or plastic pipettes range in size to contain and deliver from 0.5 to 1.0 milliliter (ml) up to 25 ml. The smallest volume measured accurately with a 1-ml pipette of this style is 0.1 ml. It is important to use a plastic bulb or other mechanical device to draw and dispense liquid from a pipette. **Never use your mouth to pipette.**

Frequently in the biotechnology laboratory, the desired volumes of liquids are smaller than 1 ml or even 0.1 ml (Fig. 2.1). Micropipettes have been developed that measure these volumes. They can accurately measure 1-µl volumes and occasionally smaller amounts. For greatest accuracy, micropipettes measure volumes within a specific range, with the micropipette measuring the smallest volumes having the smallest range. Three micropipettes commonly seen in the biotechnology laboratory have the following measuring ranges, but micropipettes with other measuring ranges are available:

- 0.5 to 10 µl
- 10 to 100 µl
- 100 to 1,000 µl

An essential component of the micropipette is a sterile pipette tip that fits on the end. The tips are hollow and hold the liquid that is being measured. Before use, the tips are sterilized to ensure that only the desired reagents are pipetted into an experiment.

Micropipettes function by depressing a plunger that displaces the air in the pipette tip. When the tip is placed beneath the surface of the liquid and the plunger

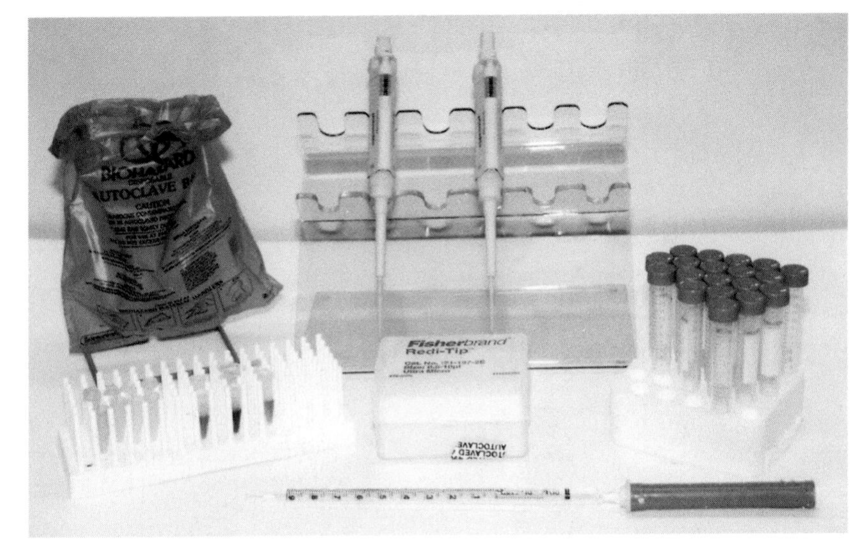

Figure 2.1 Equipment used in working with small volumes of reagents and cultures. Front, from left to right: microcentrifuge tube rack containing 1.5-ml microcentrifuge tubes, disposable tips for a 0.1- to 1.0-μl micropipette, a 10-ml serological pipette inserted into a pipette aid, and sterile 15-ml conical centrifuge tubes. Rear, from left to right: bag to collect biohazardous waste, micropipette rack holding 100- to 1,000-μl and 0.1- to 10-μl digital micropipettes.

is released, liquid is pulled into the tip. This is similar to squeezing an eye dropper bulb and then releasing it. For accuracy, the plunger reaches a point of resistance that indicates that the desired volume will be measured accurately. To dispense the liquid completely, the plunger is depressed to a second point of resistance, which releases the liquid plus additional air. Finally, the used tip is ejected. Depending on the style of the micropipette, ejection may be carried out by pushing the plunger down to a third point of resistance, or, alternatively, some micropipettes have a separate tip ejection button. Either ejection method allows one to discard the tip without touching it.

Balances

Balances are used in the laboratory to weigh reagents for the preparation of solutions (Fig. 2.2). They may also vary in the range of weights that they measure accurately, so it is important to read the instruction manual to become familiar with the specific balance used in the laboratory. Reagents are measured by placing them on a piece of weighing paper, in a weigh boat, or in a beaker. To obtain the correct amount of reagent, one must tare the container by weighing the container first and then subtracting its weight from the final reading, or tare the balance before use. Some balances have a tare that resets the balance to zero so that one may read the desired weight directly.

Centrifuges

Centrifuges (Fig. 2.2) are instruments that are used to separate cells, macromolecules such as DNA, RNA, and proteins, or subcellular organelles including mitochondria, ribosomes, and nuclei, from homogenates of cells. They may be used either preparatively for the isolation of pure components or analytically to determine sedimentation rate. We will be using them preparatively.

Preparative centrifuges vary in the sample size they will hold and in the centrifugal force they can generate. Force (g) is determined by the following formula:

$$g \text{ force} = 11.18r\,(\text{rpm}/1{,}000)^2$$

where r is the radius of the centrifuge and rpm is revolutions per minute.

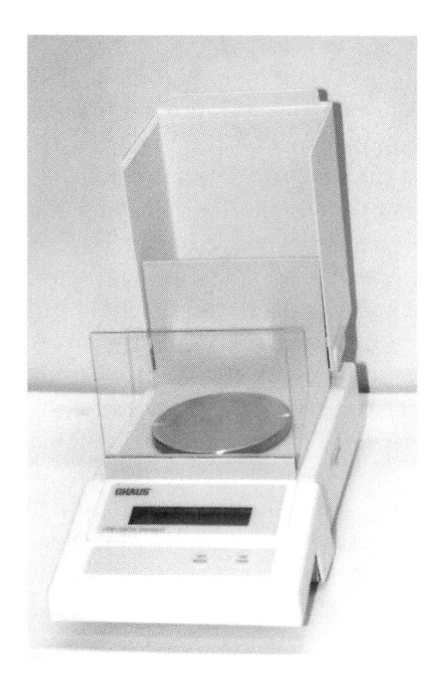

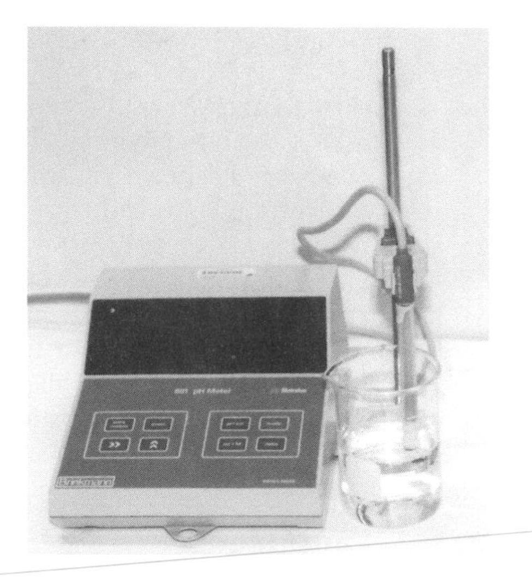

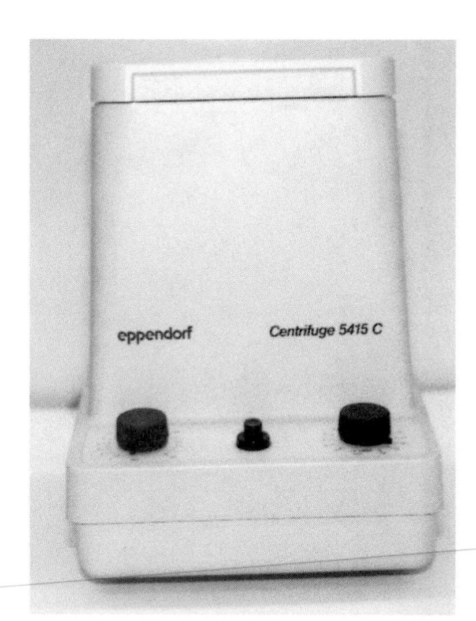

Figure 2.2 Commonly used laboratory equipment. From left to right: balance, pH meter, and microcentrifuge.

Two or more centrifuges may be present in the laboratory: a low-speed centrifuge and a microcentrifuge. The low-speed centrifuge has a fixed angle rotor and holds a maximum tube size of 15 ml. The low-speed centrifuge is capable of 4,000 rpm and is not adjustable. The microcentrifuge has a rotor that will hold 1.5-ml microcentrifuge tubes. The speed may be adjusted from 1,000 rpm to a maximum speed of 14,000 rpm.

pH Meter

The acidity or basicity of solutions is an important consideration in the preparation of buffers, growth media, and other solutions. Many biotechnology experiments and protocols rely on enzymes or reactions that occur at physiological conditions, often including a specific pH. The pH of solutions or reaction mixtures may need to be regulated or adjusted accordingly. The pH scale ranges from 0 to 14. A pH of 7 is neutral; values below 7 are acidic and those above 7 are basic. Each pH unit represents a 10-fold change in hydrogen ion concentration. There are several ways to measure and adjust pH, including colorimetrically using pH paper and electronically using a pH meter (Fig. 2.2).

pH paper consists of paper impregnated with a pH-sensitive dye. A strip of pH paper is dipped into a solution, the paper changes color, and the color change is compared with a standard. This is an acceptable and inexpensive method for determining pH, and it may be the desirable method if the volume of solution is very small.

A very sensitive method for determining and adjusting pH is with a pH meter (Fig. 2.2). A pH meter consists of an electrode and a meter. The meter provides a gauge or readout of the pH that is being measured by the electrode. Most electrodes found in modern biotechnology laboratories are combination electrodes, incorporating a glass electrode and a reference electrode. The glass electrode is permeable to hydrogen ions; thus an electrochemical potential develops across the glass. The pH meter measures the potential that develops between the glass electrode and the reference electrode.

The measurement of potential and pH is temperature dependent. Most modern pH meters include a temperature probe along with the glass and reference electrodes to automatically adjust for this variable.

Temperature Control

Additional equipment found in the biotechnology laboratory may include water baths, incubators, and heating blocks (Fig. 2.3). A common feature of these pieces of equipment is that they can be used to control the proper incubation temperature required for some experiments. One must allow the equipment to equilibrate at the correct temperature before use. If the apparatus has never been used, it should be set up and allowed to equilibrate overnight. When changing the temperature of equipment, always use a thermometer to check accuracy; the dial serves only as a rough gauge. The temperature should remain stable for at least 15 minutes before use. Water baths may be "micro," holding about 1 liter of water, or large enough to contain 10 liters or more. The smaller water baths are extremely useful for heating microcentrifuge tube reaction mixtures. Large water baths may be used to maintain appropriate temperatures in large culture vessels. Some water baths are equipped with trays or frames to provide shaking for experiments. When using a water bath, always fill it before use and make sure that it contains enough water for the entire incubation period. This is important when temperatures are significantly above room temperature and evaporation occurs.

Lab Language

accuracy
balance
centrifugation
g force
microcentrifuge
micropipette
pipette
pop spin
pulse spin
revolutions per minute
tare

Laboratory Safety

Review the sections on physical, biological, and chemical hazards in "Laboratory Safety." Special considerations for this exercise:

use of electrical equipment

Figure 2.3 Commonly used laboratory equipment. (Left) incubator; (middle and right) water baths.

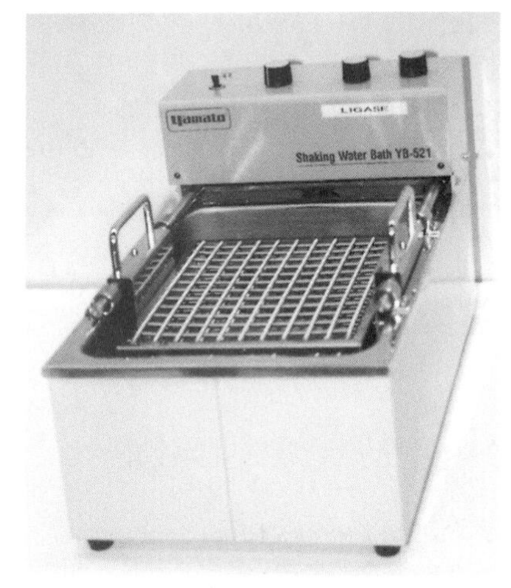

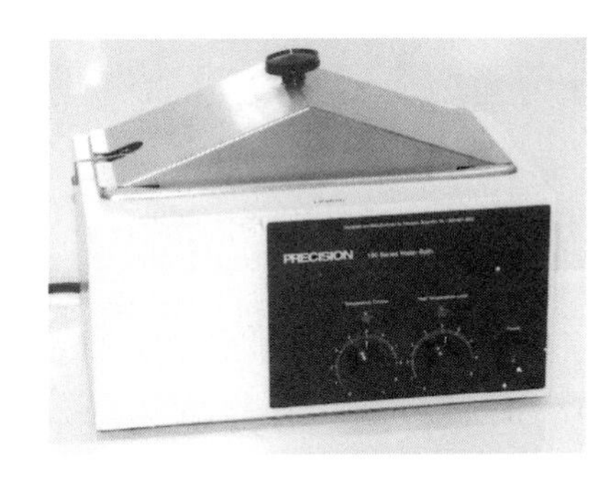

Materials

Class equipment and supplies
microcentrifuge
pH meter
balance
vinegar
household ammonia
lemon juice
10% baking soda
wash bottle with distilled H_2O

Materials per pair of students
0.5- to 10-µl micropipette and sterile tips
10- to 100-µl micropipette and sterile tips
100- to 1,000-µl micropipette and sterile tips
food coloring solutions: red, yellow, green, and blue
microcentrifuge tubes
50-ml beakers

Protocols

PROTOCOL 1 USE OF MICROPIPETTES

1. Use the micropipette only in the correct range.
2. Gently pull out the plunger and turn it to change the amount you wish to measure. Some micropipette volumes are changed by turning adjustable rings located beneath the plunger in the body of the micropipette. Consult the product information literature for your instrument. Carefully observe the dial change to avoid setting the micropipette outside its range.

A comma, bar line, or change in number color on the digital dial indicates the decimal point.

3. Use the correct size tip for each micropipette. Insert the pipette into the top of the pipette tip.

Many micropipettes are color-coded to match the correct tip.

4. The plunger stops at two or three places of resistance when depressed.
   - Depress the plunger to the first stop to measure the desired amount of liquid.
   - Depress the plunger to the second stop to dispense all of the liquid.
   - To eject the tip, depress the plunger to the third stop, or depress the tip ejection button.

Some pipettes have a separate button to eject the tip.

5. Push down to the first stop on the plunger *before* placing the pipette tip under the surface of the liquid. Release the plunger to draw liquid into the tip.
6. Transfer the liquid to a new tube by depressing the plunger to the second stop.
7. Eject the tip.
8. Practice pipetting with all micropipettes.

PROTOCOL 2 USE OF THE MICROCENTRIFUGE AND MICROPIPETTES

1. Using the 0.5- to 10-µl micropipette, dispense 5 µl of blue solution into a microcentrifuge tube. With a clean tip, dispense 5 µl of red solution into the same tube.
2. In a second tube, similarly dispense 5 µl of yellow solution and 5 µl of green solution.
3. Place the tubes in the microcentrifuge, opposite one another for balance.
4. Pulse (pop) spin the tubes by centrifuging at full speed for a few seconds.

A pulse spin may also be termed a pop spin.

5. Wait until the centrifuge has come to a complete stop and remove the tubes.

EXERCISE 2

6. Set the micropipette for 1 µl and remove the solution.

You should be able to remove 1 µl 10 times.

7. Using the 10- to 100-µl micropipette, dispense 50 µl of blue solution and 50 µl of red solution into a microcentrifuge tube as in step 1. Similarly, dispense 50 µl of yellow solution and 50 µl of green solution as in step 2.

8. Place the tubes in the microcentrifuge, opposite one another for balance.

9. Pulse (pop) spin the tubes as in steps 4 and 5.

10. Set the micropipette for 10 µl and remove the solution.

You should be able to remove 10 µl 10 times.

11. Using the 100- to 1,000-µl micropipette, dispense 500 µl of blue solution and 500 µl of red solution into a microcentrifuge tube as in step 1. Similarly, dispense 500 µl of yellow solution and 500 µl of green solution as in step 2.

12. Place the tubes in the microcentrifuge, opposite one another for balance.

13. Pulse (pop) spin the tubes as in steps 4 and 5.

14. Set the micropipette for 100 µl and remove the solution.

You should be able to remove 100 µl 10 times.

PROTOCOL 3 USE OF THE BALANCE

1. Label 10 microcentrifuge tubes 1 through 10.

2. Determine the weight of each tube.

3. Pipette 100 µl of water into tubes 1 to 5 using the 10- to 100-µl micropipette.

4. Pipette 100 µl of water into tubes 6 to 10 using the 100- to 1,000-µl micropipette.

5. Determine the weight of the tubes, now containing water.

6. Determine the weight of the water in each tube.

Weight of the water is determined by subtracting the weight of the empty tube from the weight of the tube with liquid.

Some balances have a "tare" button which zeros the balance.

Tube number	Weight of tube	Weight with water	Weight of water
1			
2			
3			
4			
5			
6			
7			
8			
9			
10			

PROTOCOL 4 USE OF THE pH METER

1. Obtain a series of reference buffers at room temperature.

Reference buffers are usually at pH 4.0, 7.0, and 10.0.

2. Plug in the pH meter and turn it on.

Consult the user's manual for your meter to determine whether you must manually adjust for the temperature of the room.

3. Place pH 4.0 reference buffer in a beaker, sufficient to cover the bottom 2 cm of the electrode.

4. Immerse the electrode in the buffer. Allow the meter to equilibrate, then adjust the meter to read 4.0.

Read the label on the reference buffer to account for temperature when adjusting the pH meter setting.

5. Remove and rinse the electrode with a stream of distilled water from a wash bottle.

6. Repeat steps 3 and 4 with the pH 7.0 buffer. Rinse as in step 5.

7. Repeat steps 3 and 4 with the pH 10.0 buffer. Rinse as in step 5.

8. Obtain a chemical or biological sample whose pH is unknown. Place enough of the sample in a clean beaker to cover the bottom 2 cm of the electrode.

Your instructor will provide a variety of samples, including vinegar, household ammonia, lemon juice, and 10% baking soda.

9. Immerse the electrode in the sample and allow the meter to equilibrate.

Be sure to rinse the electrode between each sample.

10. Read the pH and write your results in the table below.

11. Rinse as in step 5.

Sample	pH

12. When the electrode is not in use, immerse it in pH 7.0 buffer or 3.0 M KCl.

Reference

Wandersee, J. H., D. R. Wissing, and C. T. Lange (ed.). 1996. *Bioinstrumentation: Tools for Understanding Life.* National Association of Biology Teachers, Reston, Va.

Discussion

1. What is the weight of 100 µl of water as determined in protocol 3?
2. Is this determination accurate? Why or why not?
3. Which micropipette was most precise? Why do you think this is so?
4. Describe the process of centrifugation.
5. Describe the separation of two different organelles, each with different mass, from a cell lysate.
6. How does a pH meter function?
7. Discuss the pH values of any two samples you tested.

EXERCISE 3

Preparation of Solutions

Key Concepts

1. Preparation of reagents and solutions is a critical part of any experimental procedure.
2. Some materials used in a laboratory may be hazardous if handled inappropriately. Material safety data sheets should be obtained for all chemicals and consulted before use of those substances. A file or binder containing material safety data sheets should be available in the laboratory.
3. Many solutions used in the laboratory are made and stored as concentrated stock solutions but will be used at a more dilute concentration. One must be able to calculate the amount of concentrated stock reagent to obtain the final concentration and volume required.
4. Formulations for a reagent are often given for a 1-liter volume. One must be able to calculate quantities necessary to prepare a desired volume.

Goals

By the end of this exercise, you will be able to

- locate material safety data sheets in the laboratory
- prepare solutions for use in your experiments
- understand and calculate molar and normal solutions
- understand and use the "×" designation
- understand and determine percent solutions
- perform calculations to convert a standard formulation into the amounts required for any desired amount of a reagent
- determine how much stock reagent of a given concentration to use to obtain the desired final concentration

Introduction

The preparation of reagents and solutions is one of the most critical parts of any experiment. If solutions are not made properly, the experiment will not give the predicted results. Before preparing any solution, one should be familiar with the individual components, any associated hazards, and proper handling protocols.

It is important to understand the function of each component of a solution. This will allow you to determine how much precision is necessary for reagent preparation. Occasionally, less precision is required and one might "cut corners." If you're not sure, be precise. When measuring volumes, graduated cylinders are accurate. Volumes on bottles and beakers are only approximate (Fig. 3.1). Make sure you use the appropriate graduated cylinder for the job. This means one with a total volume as close to what you are measuring as possible. Do not use a 500-ml or 1,000-ml cylinder to make a solution with a total volume of 100 ml.

Figure 3.1 Laboratory equipment used for preparing solutions. From left to right: collection bag for biohazardous waste, media bottles, test tubes in foam rack, Erlenmeyer flasks, graduated cylinders, and beakers.

The Metric System

Scientists use the metric system of measurement. However, since the field of biotechnology routinely uses "micro" amounts of DNA for cloning but "macro" amounts of cultures in fermentation reactions, it is useful to review the metric system and its prefixes. Typical measurements in the biotechnology laboratory include mass, length, and volume. Mass is measured in grams (g), length is measured in meters (m), and volume is measured in liters (l).

Prefixes added to the basic unit—gram, meter, or liter—describe larger or smaller portions. The table below lists common prefixes and their relation to the basic unit of measurement.

Prefix	Arithmetic notation	Scientific notation
Kilo (k)	1,000	1×10^{3}
Milli (m)	0.001	1×10^{-3}
Micro (μ)	0.000001	1×10^{-6}
Nano (n)	0.000000001	1×10^{-9}
Pico (p)	0.000000000001	1×10^{-12}

Amounts less than microliter or milligram quantities are best obtained by making a more concentrated stock, one in which the amounts may be measured accurately, and then performing serial dilutions until the appropriate concentration is obtained.

The "×" System

Many reagents and solutions can be made in advance and stored properly for future use. Some solutions can be stored for a year or more. This can make the task of setting up and preparing your laboratory experiments much easier and more reproducible. Many items, such as buffers and antibiotics, are made as stock solutions. These stock solutions are often made at a concentration 10 times (10×) or 100 times (100×) greater than the concentration used. This allows one to easily make up a large quantity of reagent and store it efficiently. Concentrated stocks are usually more stable for longer periods of time than are dilute stocks.

Molarity

The concentration of solutions in the laboratory is frequently defined by the molar concentration of the chemicals that compose the solution. A 1 molar (1 M) solution indicates that 1 mole, that is, one formula weight, the molecular weight in grams, of the reagent is dissolved in 1 liter of solvent. The molecular weight is the sum of the atomic weights of all atoms in a molecule.

Percent Solutions

Another way that the concentration of a solution is defined is by percent. Percent means "portion of 100." Therefore, to make 100 ml of a 10% liquid solution, 0.10 × 100 ml (i.e., 10 ml) of the 100% solution is diluted to 100 ml total. This is a volume per volume dilution. When a powdered reagent is dissolved to make a liquid solution, the dilution is weight per volume and is based on 1 ml of water (the most typical solvent) weighing 1 g. For example, preparing 100 ml of 10% NaCl requires dissolving 10 g of the solid reagent in 100 ml total volume of distilled water (dH_2O). Molecular weight is not considered in this calculation.

Dilution of Stock Solutions

Once a concentrated stock is made, it must be diluted appropriately for use. When solutions are a 10× or 100× stock, this is easy. The reagent is diluted 1:10 or 1:100. But what if you don't need 10 ml or 100 ml? What if you need 75 ml or 225 ml? What if your reagent is ampicillin stored as a 10-mg/ml stock solution and you need 250 ml of Luria-Bertani (LB) agar with a final concentration of 50 μg/ml? How do you perform your calculation? Do you need to take into account the volume of the stock reagent that you are adding to your diluent? For the conversion, use the formula:

$$(\text{Concentration 1})(\text{Volume 1}) = (\text{Concentration 2})(\text{Volume 2}),$$
$$\text{or } C1V1 = C2V2$$

Another way of thinking of this calculation is: *You have = You need.* In the example above, you have 10 mg of ampicillin stock solution per ml. You need a final concentration of 50 μg of ampicillin per ml in 250 ml of LB agar. Your equation becomes:

$$(10 \text{ mg/ml})\ V1 = (50\ \mu\text{g/ml})(250 \text{ ml})$$

First convert both sides of the equation to the same units:

$$(10 \text{ mg/ml})\ V1 = (0.050 \text{ mg/ml})(250 \text{ ml}) \text{ or}$$

$$(10{,}000\ \mu\text{g/ml})\ V1 = (50\ \mu\text{g/ml})(250 \text{ ml})$$

Now solve for V1. V1 = 1.25 ml. This is the volume of ampicillin stock solution required for a final concentration of 50 μg of ampicillin per ml in 250 ml of LB agar.

Must you take the volume of added ampicillin into account? Not in this case, because the change of volume is not significant (less than 1%), and ampicillin is usually effective in concentrations from 25 μg/ml to 200 μg/ml.

Another example: You need 450 ml of 1× Tris-borate-EDTA buffer (TBE). You have a 10× stock solution of TBE. C1V1= C2V2 becomes: (10× TBE) V1 = (1× TBE)(450 ml). Solving for V1, you would measure 45 ml of the 10× TBE stock and dilute it in 405 ml of dH_2O.

In this case, you must take into account the volume of concentrate. The change of volume is 10%, which is significant. The amount of diluent to add then becomes 450 ml − 45 ml = 405 ml. The simplest way is to measure 45 ml of 10× TBE in a 50-ml cylinder, pour it into a 500-ml graduated cylinder, and then add dH_2O to adjust the volume to 450 ml total. This is also called q.s. for the Latin *quantum sufficit,* which means "enough."

Make sure that the units cancel each other out. Once an answer has been obtained, you should be able to check your calculation by plugging your answer into the variable for which you were solving.

Sterilization

Sterilization refers to any process that kills all microbes, both vegetative cells and spores. Heat, such as boiling or autoclaving, is commonly used because it is relatively easy and inexpensive. An autoclave provides steam under conditions of high temperature and pressure. The higher the pressure in an autoclave, the higher the temperature of the confined steam becomes. This effectively produces higher temperatures than those obtainable by boiling. When 100°C steam is subjected to 15 pounds of pressure per square inch (psi), the resulting temperature rises to 121°C. Autoclaving under these conditions for 15 minutes kills all organisms, including endospores. Longer times might be required for large volumes or bulky items.

It is not always useful to autoclave a reagent. Sometimes the extreme heat conditions will cause important components to denature or break down. In this case, ultrafine filtration, passing the solution through filters with pore sizes of 0.22 µm or 0.45 µm, is the method frequently chosen. The filters trap and remove most bacteria. This does not eliminate viruses, as many are small enough to pass through the pores.

Lab Language

10× concentrate
100× concentrate
autoclave
concentration
diluent
molarity
percent
q.s.
reagent
solution
sterile

Laboratory Safety

Review the sections on physical, biological, and chemical hazards in "Laboratory Safety." Special considerations for this exercise:

use of electrical equipment
use of chemicals
autoclaving

Materials

Class equipment and supplies
balance
weigh boats
spatulas
autoclave
stir plate and bars (optional)
powdered NaCl
powdered Tris base
powdered boric acid
0.5 M EDTA, pH 8.0
dH_2O
Parafilm
autoclave tape
autoclave

Students will need to prepare
100 ml of 0.9% NaCl
100 ml of 10× TBE
200 ml of 5 M NaCl

Materials per pair of students
2 125-ml bottles
1 250-ml bottle
labeling tape
5-ml pipette and pipette aid
100-ml graduated cylinder
250-ml graduated cylinder
marking pen

Protocols

PROTOCOL 1 MATERIAL SAFETY DATA SHEETS

1. Locate the material safety data sheet for NaCl.
2. List potential hazards associated with NaCl.
3. What precautions should be taken when handling NaCl?
4. How should a spill of NaCl be contained?
5. Locate the material safety data sheet for ethidium bromide.
6. List potential hazards associated with ethidium bromide.
7. What precautions should be taken when handling ethidium bromide?
8. How should a spill of ethidium bromide be contained?
9. Locate the material safety data sheet for polyacrylamide.
10. List potential hazards associated with polyacrylamide.
11. What precautions should be taken when handling polyacrylamide?
12. How should a spill of polyacrylamide be contained?
13. Locate the material safety data sheet for Tris base.
14. List any potential hazards associated with Tris base.
15. What precautions should be taken when handling Tris base?
16. How should a spill of Tris base be contained?

PROTOCOL 2 PREPARATION OF 100 ml OF 0.9% NaCl

1. Determine the amount of crystalline NaCl necessary to make 100 ml of 0.9% NaCl solution. Document your calculations below.

It is important that the NaCl solution be made precisely.

2. Measure approximately 80 ml of dH_2O into a 100-ml graduated cylinder.

Volume will increase as components are added to the dH_2O.

3. Weigh the appropriate amount of NaCl.
4. Add the NaCl to the dH_2O. Seal the graduated cylinder with Parafilm and mix by inverting two or three times until the solution is fully dissolved.
5. Adjust the volume to 100 ml by adding dH_2O q.s.
6. Carefully pour the solution into a 125-ml bottle. Label the solution with your name, date of preparation, concentration, and type of reagent.
7. Label the bottle with sterility indicator tape and sterilize in the autoclave.

Autoclave tape changes color to indicate that the sterilization process was efficient.

8. Store 0.9% NaCl at room temperature. It can be stored for months.

Your instructor will demonstrate how to use the autoclave.

PROTOCOL 3 PREPARATION OF 100 ml OF 10× TBE

1. Determine the amounts of powdered Tris, powdered boric acid, and liquid 0.5 M EDTA needed to make 100 ml of 10× TBE.

Reagent	Stock	Amount of reagent for 1 liter of 10× stock	Amount of reagent for 100 ml of 10× stock
Tris base	Powder	108 g	
Boric acid	Powder	55 g	
EDTA, pH 8	0.5 M solution	40 ml	

Volume will increase as components are added to the dH_2O.

2. Measure approximately 70 ml of dH_2O and pour it into a 125-ml bottle.

EDTA serves to chelate metal ions.

3. Measure the appropriate amount of EDTA solution and add it to the dH_2O.

Shake or use a stir plate.

4. Weigh out the appropriate amount of Tris. Slowly and carefully add it to the EDTA-dH_2O mixture. Mix until the Tris is fully dissolved.
5. Weigh out the appropriate amount of boric acid. Slowly and carefully add it to the Tris-EDTA-dH_2O mixture. Mix until fully dissolved.
6. When all components are fully dissolved, pour the mixture into a 100-ml graduated cylinder. Add dH_2O until the volume reaches 100 ml.

10× TBE may be autoclaved.

7. Pour the 10× TBE solution back into the original 125-ml bottle and label it with your name, the name and concentration of the solution, and the date.
8. Store 10× TBE at room temperature. It can be stored for months.

PROTOCOL 4 PREPARATION OF 200 ml OF 5 M NaCl

1. Obtain the molecular weight of NaCl.

Clear notes of calculations provide a mechanism for error-checking in case of experimental failure.

2. Determine the amount of NaCl needed to make 200 ml of a 5 M NaCl solution. Note your calculation below:

3. Measure about 150 ml of dH_2O and pour it into a 250-ml bottle.
4. Weigh out the appropriate amount of NaCl and add it to the dH_2O.
5. Mix well using a stir plate and stir bar, or carefully shake the capped bottle.
6. When the NaCl is completely dissolved, pour the solution into a 250-ml graduated cylinder and adjust the volume to 200 ml.
7. Carefully pour the 5 M NaCl solution back into the original bottle and label with your name, the name and concentration of the solution, and the date.

5 M NaCl may be autoclaved.

8. Store 5 M NaCl at room temperature. It can be stored for months.

PROTOCOL 5 PREPARATION OF ______________(VOLUME) OF ________________ (REAGENT)

Your instructor may assign another solution or reagent for you to make.

1. Use the table and space below to document your formulation and calculations.

Reagent	Stock	Amount of reagent (g, mM, M, %) in final solution	Amount of concentrated stock to make ____ ml of ______ (reagent)

2. Will this reagent require sterilization?

3. How will this reagent be stored?

4. For how long?

General Reading

Reger, D. L., S. R. Goode, and E. E. Mercer. 1997. *Chemistry: Principles & Practice,* 2nd ed. Saunders College Publishing, Fort Worth, Tex.

Discussion

1. When preparing a solution, why do you dissolve the components in less diluent than the desired final volume of the solution?
2. What factors affect how precise measurements must be?
3. List an example of a solution or reagent for which precision is very important. Why is this so?
4. List an example of a solution or reagent for which precision is less critical. Why is this so?
5. Describe the autoclaving process.

Calculate and describe how to make the following reagents and solutions from the stocks listed below:

6. 1 liter of LB medium with 100-µg/ml ampicillin. Stocks = 1 liter of LB medium, 10-mg/ml ampicillin.
7. 100 ml of 3 M sodium acetate solution. Stock = powdered sodium acetate.
8. 250 ml of 2× SSC , 0.5% sodium dodecyl sulfate. Stocks = 20× SSC, 10% sodium dodecyl sulfate.
9. 500 ml of 1 M Tris, pH 8.0. Stock = powdered Tris base.
10. 10 ml of 10 mM Tris, 1 mM EDTA. Stocks = 1 M Tris, 0.5 M EDTA.
11. 100 µl of 50 mM Tris, 100 mM NaCl, 1 mM dithiothreitol, 1% bovine serum albumin. Stocks = 1 M Tris, 5 M NaCl, 1 M dithiothreitol, 10% bovine serum albumin.

EXERCISE 4

Preparation and Maintenance of Microbial Cultures

Key Concepts

1. Microorganisms, especially bacteria, are very important to biotechnology. They can be used to replicate large quantities of DNA and to produce desired proteins.
2. Microbes inhabit the environment all around us. Aseptic techniques are necessary to ensure that one is working only with desired microbes, not environmental contaminants.
3. The bacterial growth curve consists of a lag phase, a phase of exponential growth, a stationary phase, and a final death phase. Biotechnology experiments frequently use bacterial cultures in exponential phase.
4. Properly prepared and catalogued microbial cultures may be stored for months to years.

Goals

By the end of this exercise, you will be able to

- work with proper aseptic techniques
- propagate microbial cultures for laboratory use
- properly store microbial cultures for future use
- understand the growth stages of bacteria
- discuss the uses of bacteria in biotechnology

Introduction

Bacteria are ubiquitous (Fig. 4.1). They live within our bodies and in the environment to assist us with normal and vital functions. When working with bacteria in the laboratory, all manipulations and experiments must be performed using aseptic conditions. All pipette tips, tubes, growth media, and solutions must be sterilized (by autoclaving or an alternative method) before use. All materials must be handled correctly to ensure that only the desired bacteria are present in an experiment. Students must think carefully about maintaining aseptic conditions, not only for culture purity but also for personal safety. **Review the safety guidelines for biological hazards.**

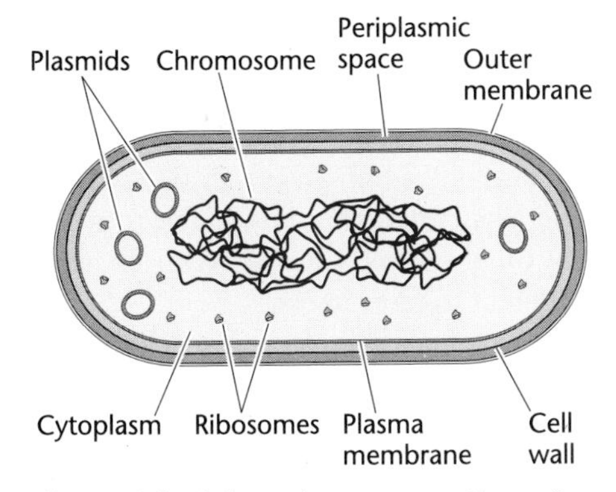

Figure 4.1 Schematic representation of a bacterial cell. (Reprinted from B. R. Glick and J. J. Pasternak, *Molecular Biotechnology: Principles and Applications of Recombinant DNA*, 2nd ed. [ASM Press, Washington, D.C., 1998], with permission.)

Aseptic techniques embody a series of manipulations designed to maintain the purity of microbial cultures. Contamination from environmental sources or from other cultures in the laboratory is thereby avoided. Proper aseptic techniques greatly reduce the potential health and safety hazards associated with working with microbes in the laboratory. Good aseptic techniques also will prevent bacterial contamination of skin, wounds, eyes, or mucous membranes. Disposable vinyl or latex exam gloves should be worn always. Pipetting by mouth is never permitted. Surfaces may be disinfected with 10% bleach or an aqueous phenol formulation. Contaminated materials must be disinfected or autoclaved before being discarded.

There is no substitute for focus, concentration, and common sense when working with microbes. **Students should view all microbes as potentially infectious.** This attitude will foster care and respect, making aseptic techniques easier to learn and consistently use.

Bacterial Growth Media

For propagation of bacteria for biotechnology experiments, proper nutrients must be provided. One commonly used liquid medium is Luria-Bertani (LB) broth. LB broth is a complex medium since the exact quantities and composition of all of the vitamins, minerals, and other nutrients are unknown and its exact chemical composition will vary from batch to batch. Other bacterial growth media may be chemically defined, which means that each component and its exact molar quantity are known and carefully measured.

The formulation for 1 liter of LB broth is:

Component	Quantity	Function
Yeast extract	5 g	Vitamins and minerals
Tryptone	10 g	Carbon, nitrogen, energy
NaCl	10 g	Proper osmotic pressure
1 M NaOH	1 ml	Proper pH

Bacterial growth media also may be solidified with agar, a complex polysaccharide derived from algae. Most microorganisms cannot digest agar, making it an ideal solidifying agent. Agar solidifies at 40 to 42°C but does not melt until the temperature is raised to about 80 to 90°C. Solid media, in petri dishes, may be used to isolate individual bacterial clones necessary to establish pure cultures and recombinant DNA libraries.

Bacterial Growth Curve

Bacteria have four well-defined growth phases: lag phase, log or exponential growth phase, stationary phase, and death phase (Fig. 4.2). In lag phase, bacteria are metabolizing nutrients but are not increasing in number. During log or exponential growth phase, bacteria are replicating at their maximum growth rate, doubling every generation. In stationary phase, nutrient limitation and toxic waste production slow bacterial growth rate. The death rate equals the reproduction rate; there is no net gain in bacterial numbers. Finally, in death phase, the detrimental environment of exhausted bacterial media can no longer sustain a healthy bacterial population. Decline in the number of viable bacteria usually is logarithmic. Log phase usually is required for biotechnology applications.

Uses of Bacteria in Biotechnology

Although *Escherichia coli* is a normal inhabitant of the human gut, strain MM294 is used commonly in academic and research labs. Apparently, it is unable to colonize the human gastrointestinal tract. Many other strains of *E. coli* have been used extensively in research labs and pose little threat to researchers.

Several characteristics of bacteria make them ideal tools in biotechnology. The commonly used bacterium, *E. coli*, doubles its population about every 20 minutes. Rapid replication coupled with small cell size allows scientists to propagate large numbers quickly in small volumes. Many bacteria possess plasmids, small circular molecules of DNA that exist extrachromosomally in the cytoplasm. Many plasmids replicate independently of the bacterium's main chromosome. These features provide molecular biologists with the opportunity to design plasmid vectors containing genes of interest in recombinant DNA technology.

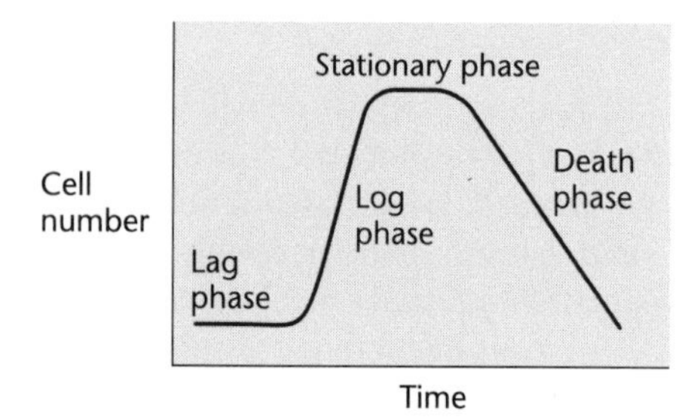

Figure 4.2 Bacterial population growth curve.

Molecular biologists have capitalized on these characteristics of naturally occurring plasmids. Using native plasmids, scientists have developed artificial plasmids constructed or recombined to contain foreign DNA (discussed in detail in exercises 13, 14, and 15). Since plasmid constructs used in biotechnology are much smaller than chromosomal DNA and replicate independently, they usually produce multiple copies once they are transformed into a suitable bacterial host cell. These features allow the biotechnologist to propagate the plasmid and the desired DNA it contains in large quantities.

Once a specific fragment of DNA or gene has been isolated into a plasmid and propagated, it has many uses. These include generation of probes for Southern blotting (exercise 16), production of recombinant protein (exercise 15), and DNA sequencing (exercise 28). In addition, plasmids are used in construction of DNA libraries (exercise 29) and in gene transfer (exercises 9, 20, and 21).

Antibiotic Use

Antibiotics are compounds that kill bacteria or inhibit their growth. They are used therapeutically to control bacterial infections such as strep throat and pneumonia. Recombinant DNA plasmids are usually constructed to contain a gene for antibiotic resistance. Ampicillin and tetracycline resistance genes are commonly employed. Once transformed, previously susceptible bacteria will display resistance to the antibiotic. This provides a mechanism to select only bacterial clones containing the gene of interest: it is on the same plasmid as the antibiotic resistance gene (see exercise 13).

Lab Language

agar
antibiotic
aseptic
autoclave
chemically defined media
colony
complex media
clone
death phase
exponential growth
lag phase
log phase
plasmid
stationary phase
sterile

Laboratory Safety

Review the sections on physical, biological, and chemical hazards in "Laboratory Safety." Special considerations for this exercise:

working with microbes
autoclaving
use of a Bunsen burner

Laboratory Sequence

Depending on the time allotted for a laboratory session, protocols 1 to 6 may be completed in 1 day; otherwise they may be allocated as follows:

Day 1: Prepare media (protocol 1)
Inoculate liquid culture (protocol 4)

Day 2: Pour agar plates (protocol 2)
Prepare glycerol stock (protocol 6)
Prepare stab culture (protocol 5)
Streak for isolated colonies (protocol 3)

Materials

Class equipment and supplies
37°C incubator
37°C shaking water bath
56°C water bath (optional)
microwave oven (optional)
autoclave
balance
weigh boats
measuring spatulas
LB broth, prepared
autoclave tape
powdered yeast extract
powdered tryptone
powdered NaCl
1 M NaOH
dH_2O
3-ml vials of LB agar
10-mg/ml ampicillin
sterile 1 M $MgSO_4$
sterile 1 M Tris, pH 8
sterile 100% glycerol
sterile dH_2O

Students will need to prepare
150 ml of LB agar
1-ml 2× glycerol storage buffer

Materials per pair of students
10- to 100-µl micropipette and sterile tips
100- to 1,000-µl micropipette and sterile tips
starter plate of bacteria
liquid starter culture of unknown bacteria
inoculating loop
inoculating needle
Bunsen burner for flaming
sterile microcentrifuge tubes
8 sterile petri dishes (17 × 100 mm)
marking pen
250-ml graduated cylinder
250-ml autoclavable bottle and cap

Protocols

PROTOCOL 1 PREPARATION OF 150 ml OF LB AGAR

1. Determine the amounts of the reagents needed to make 150 ml of LB agar.

Reagent	Stock	Amount of reagent for 1 liter	Amount of reagent for 150 ml
Yeast extract	Powder	5 g	
Tryptone	Powder	10 g	
NaCl	Powder	10 g	
NaOH	1 M	1 ml	
Agar	Powder	15 g	

Addition of the reagents will not change the volume significantly.

2. Pour 150 ml of distilled or deionized water (dH_2O) into a 250-ml Pyrex medium bottle.

Do not add agar to media and then pour it into another container; it will not transfer efficiently.

3. Weigh out the powdered reagents and add them one at a time to the dH_2O.

4. Measure the appropriate volume of 1 M NaOH solution and add to the mixture.

The agar and other reagents will fully dissolve when the medium is autoclaved.

5. Add the appropriate amount of agar to the bottle.

Plates may be poured immediately after autoclaving; however, condensation on the lids may be reduced by cooling the agar in a 50°C water bath for 30 minutes.

6. Label the bottle with the reagent name, your name, and the date.

7. Label the bottle with autoclave tape, and autoclave at 121°C and 15 pounds of pressure for 15 minutes to sterilize.

Alternatively, the agar medium may be allowed to solidify and stored for future use, then remelted in a microwave oven when needed.

PROTOCOL 2 POURING AGAR PLATES

EXERCISE 4

1. Obtain eight sterile petri dishes and label the bottoms of four dishes with "LB" and four dishes with "LB + amp" plus your name and the date.

Petri dishes (17 × 100 mm) require about 20 ml of medium each.

2. Determine the amount of 10-mg/ml (concentrated stock solution) ampicillin needed to make 75 ml of LB agar containing a final concentration of 100-μg/ml ampicillin. Document your calculations below.

3. Place the closed dishes (bottom down) on a lab bench where they may remain undisturbed until they solidify.

4. Remove the cap of the bottle, briefly flame the neck, and hold the opening at a slight downward angle to prevent unwanted airborne microbes from entering the bottle.

Plates may be poured immediately when the medium is removed from the autoclave, but condensation will form on the lids.

5. Carefully lift one side of each petri dish lid and pour liquid LB agar into the four plates labeled "LB." Gently swirl the plate so the medium covers the bottom.

6. When the remaining 75 ml of LB agar is at about 56°C, add the correct volume of ampicillin stock to obtain a final concentration of 100 μg/ml. Gently mix.

LB agar may be placed in a 56°C water bath to cool before use.

7. As in step 5, pour liquid LB agar + ampicillin into each of the four plates labeled "LB + amp." Gently swirl the plates so the medium covers the bottom.

Medium must be cooled before adding ampicillin to prevent inactivation.

8. Allow the plates to remain undisturbed on the lab bench until the agar has completely solidified.

Other antibiotics may be heat or light sensitive; appropriate care must be taken when dispensing them.

9. These plates may be used in exercise 9.

Agar plates may be stored at 4°C in plastic bags to prevent dehydration.

Plates containing antibiotics may be stored for 1 month at 4°C.

PROTOCOL 3 STREAKING CULTURES FOR ISOLATION OF COLONIES

1. Obtain a broth starter culture of unknown bacteria and an LB agar plate.

2. Label the bottom of the petri dish with your unknown number, the date, and your name. Place it upside down on the lab bench.

3. Sterilize an inoculating loop by heating it in a flame until it is red hot. *Do not put it down.*

Your instructor will demonstrate the proper method to streak for isolation.

4. Pick up the liquid culture, remove the cap, and flame the top of the tube.

5. Lift the bottom of the petri dish and cool the loop by touching it to the surface of the sterile agar.

6. Obtain a loopful of bacteria by dipping the now cooled loop into the liquid culture.

7. Lift the bottom of the petri dish and gently streak the inoculating loop back and forth across about one-third of the agar surface (Fig. 4.3).

The first application of bacteria delivers the bacteria to the agar.

8. Flame the inoculating loop again, cool it by touching it to an uninoculated portion of the agar, and then streak it once across the first pass of bacteria.

The second streaking picks up a few bacteria and spreads them out on the agar surface.

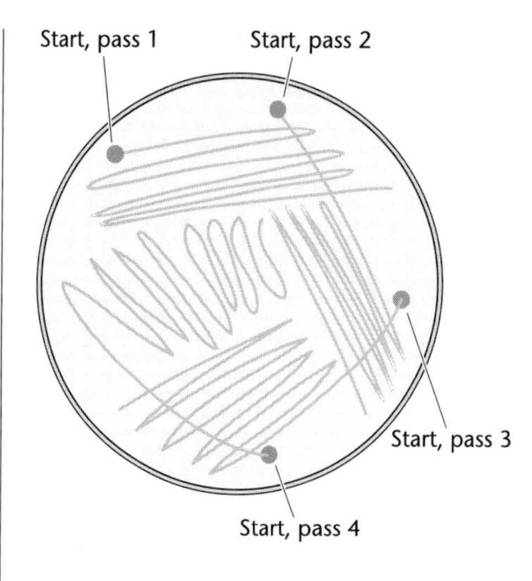

Figure 4.3 Streaking an agar plate to obtain isolated colonies (four-way-streak method). Pick up the inoculum using a flamed loop, and streak the first pass. For each subsequent pass, flame the loop before and after streaking. Overlap each pass only once or twice. Some researchers find it satisfactory to use a three-way method.

Streak gently back and forth across about another third of the sterile agar (Fig. 4.3).

The third streaking picks up even fewer bacteria, spreading them out on the agar surface so that individual colonies can be isolated.

9. Flame the inoculating loop again, cool it by touching it to an uninoculated portion of the agar, and then streak it once across the second pass of bacteria. Streak gently back and forth across the remaining uninoculated third of the agar (Fig. 4.3).

10. Replace the bottom of the dish into the lid.

11. Incubate the agar plates upside down at 37°C for 18 to 24 hours, then examine your plates for growth.

PROTOCOL 4 INOCULATION AND GROWTH OF BACTERIA IN LIQUID CULTURE

Dispose of contaminated medium in the designated contaminated-waste container.

1. Obtain a starter plate culture of bacteria and two tubes of sterile LB broth (Fig. 4.4). Check the sterile LB broth for clarity. If cloudy, the broth is contaminated.

2. Label one tube with your name, the date, and the name of the culture.

Flame as in protocol 3, above.

Using an isolated colony ensures that bacteria inoculated are clonal.

3. Flame an inoculating loop, cool it by touching it to an uninoculated portion of the agar, and then pick an isolated colony of bacteria by gently touching the loop to the surface of the colony.

4. Uncap and flame the top of the LB broth (Fig. 4.4).

5. Inoculate by mixing the loop in the LB broth.

6. Flame the top of the LB broth again and cap. Flame the inoculating loop before placing it on the lab bench.

7. Label the second LB broth tube with your name, the date, and "control."

8. Perform a control inoculation by flaming and cooling a loop as in step 3. *Do not pick a colony!* Repeat steps 4 to 6.

Shaking the bacteria culture aerates the medium for optimal bacterial growth.

9. Place both liquid cultures in a 37°C shaking water bath.

10. Incubate for 18 to 24 hours, then observe the cultures for growth.

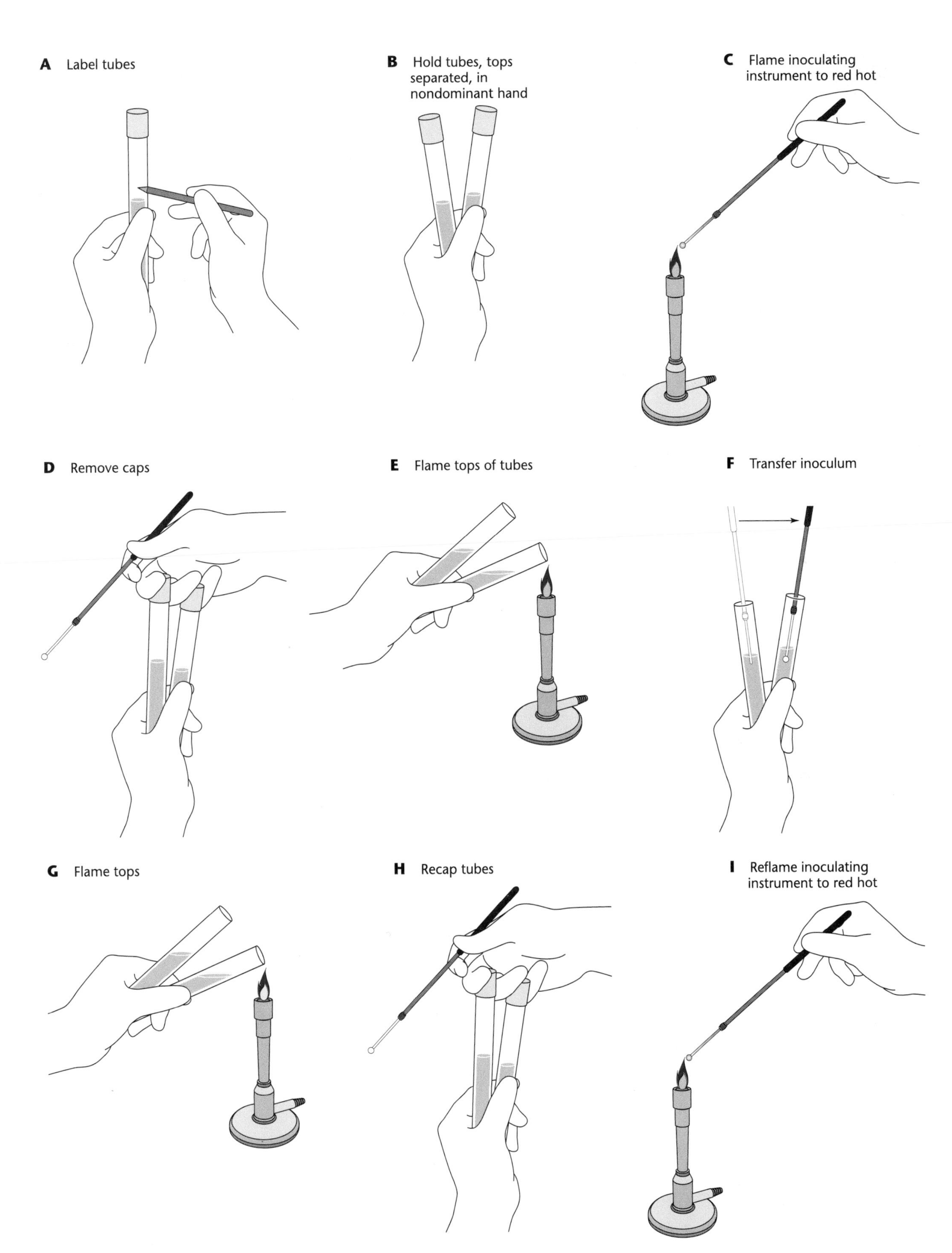

Figure 4.4 Aseptic transfer technique.

PROTOCOL 5 **INOCULATION OF A STAB CULTURE**

1. Obtain a starter plate of bacteria and a 3-ml vial of LB agar. Label the vial with your name, the date, and the name of the culture.
2. Flame an inoculating needle and touch it to an uninoculated portion of the agar to cool.
3. Touch the sterile, cooled needle to an isolated colony of bacteria.
4. Stab the needle into the agar 5 to 10 times.
5. Incubate the stab culture at 37°C for 18 to 24 hours.
6. After incubation, ascertain that growth is present.
7. Seal the stab culture with Parafilm and store it at 4°C.

PROTOCOL 6 **PREPARATION OF A GLYCEROL STOCK**

1. Obtain an overnight broth culture of bacteria in log phase.
2. Determine the amounts of reagents to make 2× glycerol storage buffer (GSB).

Reagent	Stock	Final concentration in 2× GSB	Volume needed to make 1 ml of 2× GSB
Tris, pH 8	1 M	25 mM	
$MgSO_4$	1 M	0.1 M	
Glycerol	100%	65%	
Sterile dH_2O			
Total volume			1 ml

3. Obtain a sterile 1.5-ml microcentrifuge tube. Label it with your name, the date, and the name of the culture.
4. Mix equal volumes of the overnight culture and 2× GSB in the sterile 1.5-ml microcentrifuge tube.
5. Store the glycerol stock at –20°C.

References and General Reading

Ausubel, F. M., R. Brent, R. E. Kingston, D. D. Moore, J. G. Seidman, J. A. Smith, and K. Struhl (ed.). 1994. *Current Protocols in Molecular Biology.* Wiley Interscience, New York, N.Y.

Cappuccino, J. G., and N. Sherman. 1998. *Microbiology: a Laboratory Manual,* 5th ed. Benjamin/Cummings, Menlo Park, Calif.

Discussion

1. What is agar, and why is it used in bacterial culture media?
2. Differentiate between complex media and defined media.
3. Which components of LB medium must be measured carefully? Why?

4. Which components of LB medium do not need to be measured as carefully? Why?
5. Why must you use an isolated colony to start a bacterial culture?
6. Why are most bacteria incubated at 37°C?
7. When are bacteria grown at temperatures other than 37°C?
8. Why are bacteria used in biotechnology?
9. Draw your plate that was streaked for isolation of individual colonies.
10. Discuss the patterns of growth obtained within each streak area.
11. Were individual colonies obtained? Why or why not?
12. Compare the control and inoculated liquid cultures. What do you observe? Explain.
13. Describe the stab culture after growth.
14. What are the final concentrations of GSB once an equal volume of the overnight bacterial culture is added?
15. Discuss at least three biotechnology applications in which bacteria are used.

EXERCISE 6

Investigating the Structure and Bond Strength of DNA

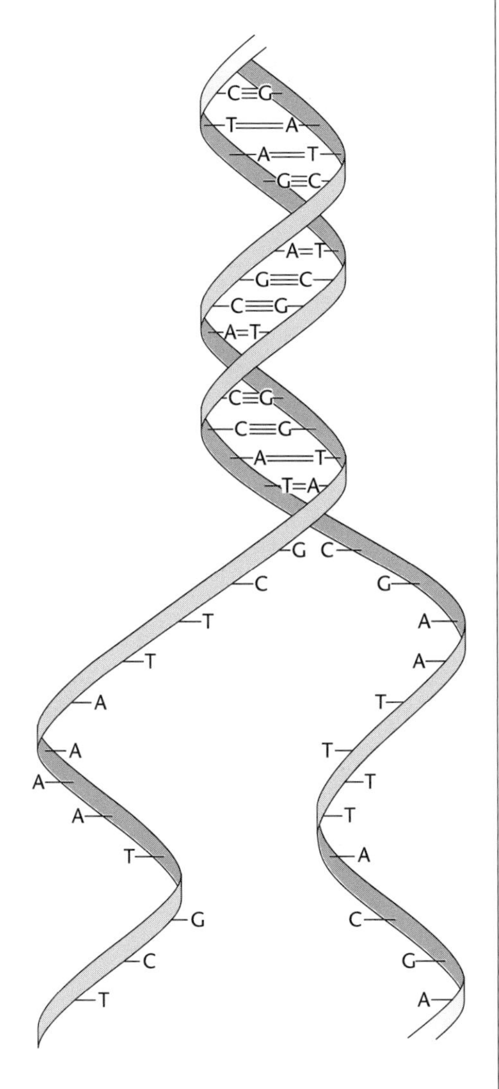

Figure 6.1 Partially denatured DNA. Double-stranded region shows G-C base pairs (three hydrogen bonds) and A-T base pairs (two hydrogen bonds). Single-stranded region shows loss of base-pairing; all hydrogen bonds are broken.

Key Concepts

1. Ultraviolet (UV) spectroscopy can be used to analyze the structure of DNA.
2. The temperature at which double-stranded DNA is denatured into single-stranded DNA is known as the melting temperature (T_m).
3. An increase in absorbance at 260 nm (A_{260}) occurs when double-stranded DNA is denatured to single strands. This is referred to as the hyperchromic effect.
4. Acridine orange and a UV light source can be used to detect double- and single-stranded DNA.

Goals

By the end of this exercise, you will be able to

- isolate a sample of DNA
- determine the absorbance change at 260 nm in heated samples of DNA
- plot a melting curve for a DNA sample
- identify the T_m on a melting curve of DNA
- use acridine orange to detect double- and single-stranded DNA

Introduction

Knowledge of DNA's unique chemistry and double-helical structure is important to researchers' ability to manipulate the molecule. Polymerase chain reaction, for example, is dependent upon changing double-stranded DNA structure into a denatured single-stranded form (Fig. 6.1).

DNA structure can be studied by utilizing well-known information about the molecule's behavior at temperatures near 100°C. When double-stranded DNA is heated, hydrogen bonds (Fig. 6.2) between the nitrogenous bases are disrupted and the complementary strands dissociate into single strands. The molecule is denatured. The temperature at which the strands separate is known as the melting temperature (T_m). The specific T_m depends on the base pair composition of DNA and the relative proportion of guanine-cytosine (G-C) and adenine-thymine (A-T) pairs. The three hydrogen bonds linking G-C pairs enhance the stability of a G-C-rich molecule and raise its T_m compared with a DNA sample rich in A-T pairs linked with two hydrogen bonds. The overall length of a DNA molecule affects its T_m. A longer fragment of DNA will contain more hydrogen bonds than a shorter fragment of DNA and therefore will be more difficult to denature.

The T_m of DNA can be determined by spectrophotometrically measuring the absorbance of the denatured DNA sample at a wavelength of 260 nm. Relatively little UV light is absorbed when the bases are paired (double stranded). When the bases are exposed (single stranded), however, more light is absorbed. This in-

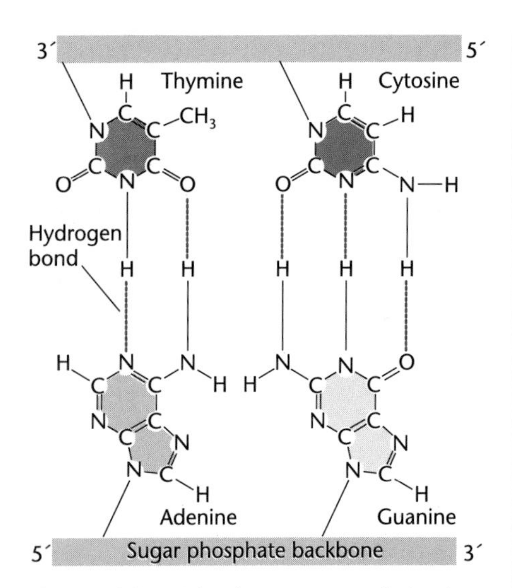

Figure 6.2 Molecular structure of the four DNA bases showing hydrogen bonding between thymidine and adenine (T and A) and guanine and cytosine (G and C).

crease in absorbance is known as the hyperchromic effect. A melting curve can be calculated by measuring the absorbance of DNA samples incubated at various temperatures and determining the T_m. The point at which the absorbance increase is one-half the maximum increase at the highest temperature is the melting point. Graphically, this is the midpoint of the transition on the curve.

Acridine orange is a fluorescent stain that is used to determine whether a DNA sample is single or double stranded. Acridine orange is added to the sample, followed by excitation with a UV transilluminator. A DNA sample fluoresces green if double stranded and orange if denatured to single strands.

Lab Language

acridine orange
bond strength
denaturation
hyperchromicity
melting temperature
renaturation

Laboratory Safety

Review the sections on physical, biological, and chemical hazards in "Laboratory Safety." Special considerations for this exercise:

- use of acridine orange
- heating of samples to temperatures near 100°C
- use of a UV transilluminator

Materials

Class equipment and supplies
liquid kitchen dishwashing detergent
UV spectrophotometer
quartz cuvettes
65°C water bath or heat block
80°C water bath or heat block
95°C water bath or heat block
UV transilluminator
ice bucket
2% acridine orange
1× SSC (15 mM sodium citrate, 150 mM NaCl, pH 8.0) buffer
20× SSC buffer

Students will need to prepare
50 ml of 50% detergent solution
0.9% NaCl (exercise 3)

Materials per pair of students
2 sterile 50-ml centrifuge tubes
2 glass rods
2 sterile 15-ml tubes
95% ethanol
1× SSC buffer
clear microcentrifuge tubes
5 quartz cuvettes
0.1- to 1.0-ml micropipette and sterile tips
0.5- to 10-µl micropipette and sterile tips
transfer pipette (5 ml)
disposable exam gloves
forceps or tongs
graph paper

Protocols

PROTOCOL 1 ISOLATION OF DNA

1. Prepare 50 ml of 50% detergent. Document your calculation below.

2. Using a sterile 50-ml tube, measure 15 ml of sterile 0.9% NaCl solution. Rinse your mouth for 60 seconds with the solution to collect cheek cells. Expel the solution, now containing the cheek cells, back into the 50-ml tube.

Cheek cells are continually sloughed from the inside of the mouth.

3. Add 15 ml of 50% detergent solution to the cell solution.
4. Gently invert the tube to mix, but *do not shake!*
5. Incubate the tube in a 65°C water bath for 10 minutes.

Heat assists with dissociation of DNA and associated proteins.

6. Using a 5-ml transfer pipette, carefully layer 15 ml of 95% ethanol on top of the cell suspension.

Alcohol causes DNA to precipitate from solution.

7. Spool the DNA onto a glass rod, being careful not to mix the layers.
8. Be careful not to disturb the DNA. Blot off the excess ethanol and air dry for a few minutes.
9. Dissolve the DNA in 10 ml of 1× SSC buffer.

PROTOCOL 2 DENATURATION OF THE DNA

1. Place 1-ml DNA samples into four separate clear tubes labeled A, B, C, and D.
2. Incubate each tube for 10 minutes at a different temperature: room temperature, 65°C, 80°C, 95°C, or other temperatures of your choice.

A more accurate melting point determination can be made by using additional incubation temperatures.

3. Record your temperature choices in the table in protocol 3.
4. Remove the samples from the water baths and place immediately in ice.

Samples must be iced to prevent renaturation.

PROTOCOL 3 MEASURING THE A_{260} CHANGES—THE HYPERCHROMIC EFFECT

1. Place 1 ml of 1× SSC buffer in a quartz cuvette and insert the cuvette into the spectrophotometer set at 260 nm. Zero or blank the instrument.

To zero or blank the instrument means to set the absorbance to zero.

2. Pipette 1 ml of each sample, incubated at different temperatures, into four separate cuvettes.

Using 1× SSC to zero the instrument ensures that sample absorbance readings will be due to DNA.

3. Read each sample at 260 nm and record the absorbance values.

Tube	Temperature	Absorbance	Color
A			
B			
C			
D			

4. Pour the samples back into the tubes for protocol 4, below. Return the samples to ice.

Icing the samples will prevent renaturation.

5. Construct a melting curve of your sample by plotting the absorbance values on the *y* axis and the temperature on the *x* axis.
6. Determine the exact T_m of your sample. The temperature of the midpoint of the transition is the melting point. Record your result.

Samples must be in clear tubes.

Green indicates double-stranded DNA, and orange indicates single-stranded DNA.

PROTOCOL 4 ACRIDINE ORANGE TEST FOR STRANDEDNESS

1. Add 4 µl of 2% acridine orange solution to each tube. Mix by inverting the tubes.
2. Excite the acridine orange bound DNA by holding each tube over a UV transilluminator and record the color in the table in protocol 3.

References and General Reading

Clerc, S., and Y. Barenholz. 1998. A quantitative model for using acridine orange as a transmembrane pH gradient probe. *Anal. Biochem.* **259:**30291–30295.

Darzynkiewicz, Z. 1990. Differential staining of DNA and RNA in intact cells and isolated cell nuclei with acridine orange. *Methods Cell Biol.* **33:**285–288.

Darzynkiewicz, Z., and J. Kapuscinski. 1990. Acridine orange: a versatile probe of nucleic acids and other cell constituents, p. 291–314. *In* M. R. Melamed, T. Lindmo, and M. L. Mendelsohn (ed.), *Flow Cytometry Sorting*, 2nd ed. John Wiley & Sons, New York, N.Y.

Freifelder, D. 1982. *Physical Biochemistry.* W. H. Freeman & Co., New York, N.Y.

Lehninger, A. L., D. L. Nelson, and M. M. Cox. 1993. *Principles of Biochemistry.* Worth Publishers, New York, N.Y.

Watson, J. D., and F. H. C. Crick. 1953. A structure for deoxyribose nucleic acid. *Nature* **171:**737–738.

Watson, J. D., N. H. Hopkins, J. W. Roberts, J. A. Steitz, and A. M. Weiner. 1987. *Molecular Biology of the Gene*, 4th ed. Benjamin/Cummings, Menlo Park, Calif.

Discussion

1. Discuss the factors that stabilize the DNA molecule.
2. Describe one or more technologies that rely on an understanding of nucleic acid chemistry.
3. What effect did heating have on the absorbance of the DNA sample? Explain.
4. Different samples of DNA may have a different T_m. Explain why this could occur.

SECTION II

The Techniques

EXERCISE 7

Restriction Enzyme Analysis of Lambda DNA

Key Concepts

1. Restriction enzymes are proteins synthesized by and obtained from bacteria. They cleave DNA at specific palindromic nucleotide sequences.
2. Restriction enzymes probably evolved in bacteria for protection against bacteriophage infection. In nature, along with each restriction enzyme, bacteria produce a corresponding methylase.
3. Restriction enzymes with over 200 different specificities have been isolated. They are essential in recombinant DNA technology, enabling scientists to sort out, manipulate, and recombine DNA molecules.
4. Gel electrophoresis is the process that is used to separate DNA fragments based on their length.

Goals

By the end of this exercise, you will be able to

- understand how restriction enzymes function
- discuss gel electrophoresis as a technique for sorting out DNA fragments of various lengths
- discuss the differences in DNA fragment bands produced by cutting lambda (λ) bacteriophage DNA with three different restriction enzymes
- understand how restriction analysis is fundamental to recombinant DNA technology

Introduction

Restriction enzymes were first isolated in 1962 by Werner Arber. His work was based on Salvador Luria's observations of bacteriophage infection cycles in the 1950s. Luria noted that some *Escherichia coli* strains were not susceptible to phage infection. The bacteriophages had a restricted host range and could only reproduce in certain strains of bacteria. Using extracts from *E. coli*, Arber's research group showed that a bacterial enzyme cut the phage DNA into fragments, thus disrupting viral replication. They called this first enzyme an endonuclease. Now these enzymes are called restriction endonucleases or restriction enzymes because they restrict viral replication and only recognize site-specific or restricted sequences of DNA.

Arber's first endonuclease was not useful for manipulating DNA, however. It did not cut DNA in reproducible, predictable places. In 1970, Hamilton Smith's laboratory isolated a new restriction enzyme from *Haemophilus influenzae*, *Hin*dIII. Then Daniel Nathan showed that *Hin*dIII cut simian virus 40 DNA at the same location in every trial. *Hin*dIII was a precise, specific, and reproducibly acting enzyme. This demonstrated the predictability and, therefore, the usefulness of

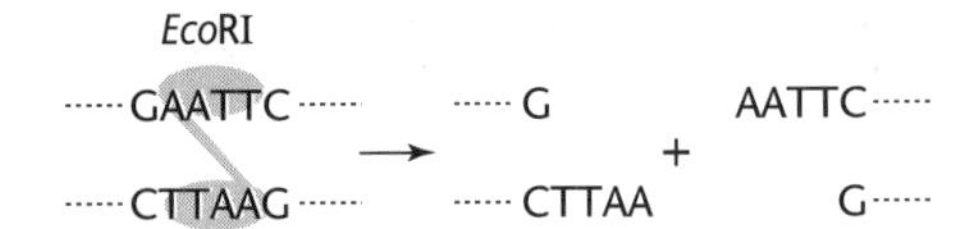

Figure 7.1 Unmodified DNA is cleaved by the restriction enzyme *Eco*RI.

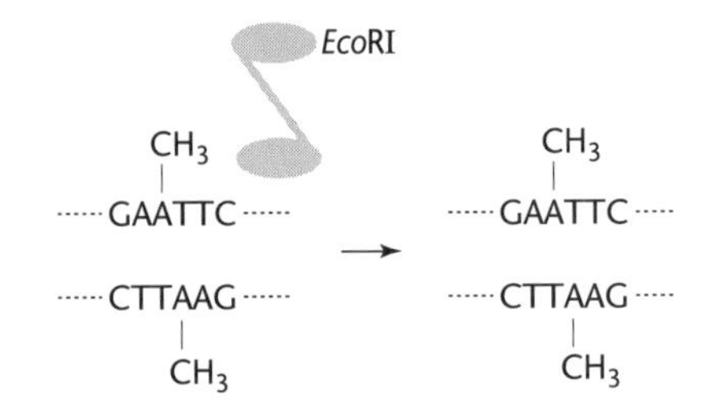

Figure 7.2 *Eco*RI is unable to cleave the methylated DNA.

restriction enzymes. Arber, Smith, and Nathan were awarded the Nobel Prize in physiology or medicine in 1978 for their work.

Bacteria lack a nucleus. Chromosomal DNA is concentrated in the nucleoid, an irregularly shaped region in the cell. The DNA, therefore, contacts the cytoplasm. This means that bacteria must protect their own DNA from the restriction enzymes produced by the adjacent cytoplasmic ribosomes. A system of DNA modification evolved, altering the restriction enzyme binding site. Modification is accomplished by a corresponding methylase, an enzyme that adds methyl (CH_3) groups to the site-specific DNA sequence recognized by the restriction enzyme. When a restriction enzyme is produced and encounters the chromosomal DNA, it can no longer recognize the modified binding site (Fig. 7.1 and 7.2). The bacterium's own DNA is protected from restriction enzyme cleavage.

Restriction enzymes are homodimers: two identical protein subunits linked together. This structure allows the enzyme to bind to its palindromic sequence (Fig. 7.1) and cut the phosphodiester bonds between the appropriate nucleotides in each strand simultaneously. A palindromic sequence is one in which the complementary sequence reads the same except backward; for example, the palindrome of GAATTC is CTTAAG.

Three different types of restriction enzyme classes have been defined. Type I restriction enzymes recognize a specific DNA sequence but cleave the DNA randomly. Type II restriction enzymes recognize a specific DNA sequence and cut that sequence in a predictable and specific manner. Type III restriction enzymes recognize a specific sequence and then cut the DNA at a location 20 to 25 nucleotides away from the recognition sequence. Type II restriction enzymes are the most useful. Very few type I or type III enzymes have been isolated.

Type II restriction enzymes recognize a palindromic nucleotide sequence in DNA and cut the DNA at that site (Fig. 7.3). Over 2,750 type II restriction enzymes have been isolated, encompassing 211 different specificities. About 15 to 20 new enzymes with different specificities are discovered each year.

A few restriction enzymes and their recognition sequences are listed below. Recognition sequences vary in length, usually 4 base pairs (bp), 6 bp, or 8 bp. Digestion may result in "blunt-ended" cuts, seen with *Sma*I, or 5′ or 3′ single-stranded overhangs, seen with *Eco*RI and *Hin*dIII, respectively (Fig. 7.3).

These short overhangs are called "complementary" or "sticky ends." The very short single-stranded base sequences that extend from the cut fragment of DNA are very useful in DNA cloning experiments. These single-stranded overhangs of DNA will bond with their complementary bases on another fragment of DNA (Fig. 7.3). This makes it possible to join DNAs from two different sources, an essential component of recombinant DNA technology.

Enzyme	Site		Resulting cut
*Eco*RI	5′ GAATTC 3′ 3′ CTTAAG 5′	→	5′ G + AATTC 3′ 3′ CTTAA G 5′
*Hin*dIII	5′ AAGCTT 3′ 3′ TTCGAA 5′	→	5′ AAGCT + T 3′ 3′ T TCGAA 5′
*Bam*HI	5′ GGATCC 3′ 3′ CCTAGG 5′	→	5′ G + GATCC 3′ 3′ CCTAG G 5′
*Sma*I	5′ CCCGGG 3′ 3′ GGGCCC 5′	→	5′ CCC + GGG 3′ 3′ GGG CCC 5′
*Hba*I	5′ GCGC 3′ 3′ CGCG 5′	→	5′ GCG + C 3′ 3′ C GCG 5′

Figure 7.3 Restriction enzyme sites recognized by various restriction enzymes and the resulting ends produced after digestion.

EXERCISE 7

DNA restriction enzyme analysis is a powerful technique that allows researchers to cut DNA into small fragments for analysis. The DNA can come from any source: humans, plants, bacteria, and other creatures. Restriction enzymes then provide researchers with the opportunity to recombine fragments of DNA from different organisms. This process is especially useful in analyzing and sequencing genes. It also provides biotechnologists with a mechanism for producing gene products for the treatment of diseases, such as insulin, inexpensively and safely.

Agarose Gel Electrophoresis

Once DNA has been cut into smaller fragments, the fragments must be separated, sorted by size, and visualized. This is performed using agarose gel electrophoresis and subsequent staining of the DNA. Electrophoresis means "carrying with electricity." DNA has a negative charge, owing to its acidic phosphate groups. Therefore, if an electric current is applied to the DNA, the DNA will move toward the positive electrode.

For separation of fragments based on size or molecular weight, the DNA sample must be suspended in an agarose gel, a semisolid matrix support. Agarose is a purified form of agar, available as a powdered reagent. When mixed with buffer and heated to boiling, it makes a homogeneous solution, which hardens and becomes semisolid at room temperature. During solidification, pores are formed in the agarose gel through which DNA fragments are forced to travel when placed in an electric current. The larger the DNA fragment, the more difficult it is to migrate through the pores in the gel and the more slowly it will travel. Smaller fragments of DNA will migrate more quickly and easily. An inverse relationship exists between the size of the DNA fragment and the distance it migrates in the agarose gel. Identically sized fragments of DNA will migrate to the same position in the gel. In this way, DNA fragments of different sizes are separated from one another.

Altering the agarose concentration changes the resolving capacity of the agarose gel. Increasing the amount of agarose decreases the pore size of the gel; larger DNA fragments move through the gel with difficulty. Smaller fragments of DNA also migrate more slowly, making it possible to resolve small DNA fragments of similar sizes. Decreasing the amount of agarose will increase the pore size of the agarose gel; small DNA fragments move quickly and uniformly through the gel. Larger DNA fragments of similar sizes will be resolved more effectively. The following table provides a guide for selection of agarose concentration based on the size of the DNA to be analyzed.

% Agarose in gel	Size of fragments resolved	Example(s)
0.5	1,000–30,000 bp	Undigested genomic DNA
0.8	800–12,000 bp	Digested genomic DNA
1.0	500–10,000 bp	Plasmid DNA
1.2	400–7,000 bp	Plasmid DNA
1.5	200–3,000 bp	Polymerase chain reaction (PCR) products and plasmids
2.0	100–1,500 bp	PCR products

Once electrophoresis is completed, the DNA must be made visible. Ethidium bromide is the most commonly used stain in research laboratories. The molecule intercalates between the DNA base pairs, analogous to slipping an extra rung into a ladder. When the ethidium bromide is excited by ultraviolet light, it emits light in the visible range and the DNA can be seen. Ethidium bromide, however, has been shown to be mutagenic by the Ames test.

Other DNA stains are nontoxic. These include methylene blue, Carolina Blu (Carolina Biological Supply), and Bio-Safe (Bio-Rad, Inc.). These stains are considered relatively safe and do not require excitation. They are not as sensitive, however. Ethidium bromide can detect as little as 10 ng of DNA. Fragments containing 100 ng of DNA are therefore easily visualized. This means that small amounts of tiny fragments of DNA may not be detected using the safer stains.

After electrophoresis and subsequent staining of the gel, the fragments of DNA will appear as bands in the gel. Each band contains thousands of DNA molecules, all of which are the same size. The array of bands forms a pattern that is sometimes called a "fingerprint."

Restriction enzyme analysis of DNA, such as you are about to perform, is often one of the first experiments to which scientists subject a newly isolated DNA segment. One must determine the fragment sizes of DNA resulting from restriction enzyme digestion and orient the DNA fragments in their correct order. Based on this information, researchers then construct a restriction enzyme map of the newly isolated DNA. This process is assisted by having a marker or DNA standard—DNA fragments of known size—which is run on the gel with the unknown sample. The size of DNA fragments is measured in base pairs. You will determine the size of the fragments of DNA generated by restriction digestion by constructing a standard curve using DNA fragments of known base pair size and then comparing your results with the standard curve.

Lab Language

agarose
aliquot
blunt-end
buffer
control
digest
DNA marker
electrophoresis
endonuclease
ethidium bromide
incubate/incubation
methylase
negative control
palindrome
positive control
procedure/protocol
solution
sticky end

Laboratory Safety

Review the sections on physical, biological, and chemical hazards in "Laboratory Safety." Special considerations for this exercise:

boiling agarose
performing gel electrophoresis
possible use of ethidium bromide

Materials

Class equipment and supplies
37°C water bath
microcentrifuge
lambda DNA
*Hin*dIII
*Bam*HI
*Eco*RI
10× restriction enzyme buffer
Carolina Blu stain concentrate
Carolina Blu final stain
0.8% agarose
10× Tris-borate-EDTA buffer (TBE) (or from exercise 3)
masking tape
dH_2O
DNA marker
white light transilluminator (optional)
Polaroid camera and black-and-white type 667 film (optional)

Students will need to prepare
300 ml of 1× TBE

Materials per pair of students
gel electrophoresis chamber, gel casting tray, comb

power supply (two pairs of students will share this)
0.5- to 10-μl micropipette and sterile tips
1 aliquot of loading dye
8 sterile microcentrifuge tubes
microcentrifuge tube rack
disposable exam gloves
permanent marking pen
beaker filled with crushed ice
1 aliquot of DNA size markers

Keep the following reagents on ice
1 aliquot of 10× restriction enzyme buffer
1 aliquot of lambda DNA (0.5 μg/μl)
1 aliquot of *Hin*dIII
1 aliquot of *Bam*HI
1 aliquot of *Eco*RI
1 aliquot of dH_2O

Protocols

PROTOCOL 1 PREPARING THE RESTRICTION ENZYME DIGESTS

1. Prepare your digests in 1.5-ml microcentrifuge tubes. In each tube, for each digest, you will need 4 μl of DNA, 1 μl of restriction enzyme buffer, and 1 μl of enzyme. You have three restriction enzymes from which to choose: *Eco*RI, *Hin*dIII, and *Bam*HI. Add one or more to each tube. Add dH_2O to bring the volume up to 10 μl.

1 unit of restriction enzyme cuts 1 μg of DNA in 1 hour.

Buffer provides the proper salt concentration for enzyme functioning.

2. Use the space below to design your experiment. You may perform up to seven different digests, *but one of them must be an experimental control.* Discuss your plan with your instructor before beginning. Reserve one lane for a DNA size marker control.

Each enzyme has its own specific recognition sequence.

Enzymes are expensive: add them last to minimize the cost of starting over in case of error.

Using a fresh tip avoids contaminating reaction mixtures with undesired enzymes.

3. Set up the digestions.

Important: First add water, then the DNA, then the restriction buffer, and then add the restriction enzymes last to the tubes. Use a fresh pipette tip for the restriction enzyme buffer and each enzyme when adding them to each of your digest tubes.

4. Securely fasten the cap on each tube. To mix the reagents, pulse the tubes in the microcentrifuge.

Pulsing the tubes mixes the reagents.

Tubes may be gently flicked with a finger and tapped on the benchtop.

5. Incubate the sample tubes in a 37°C water bath for approximately 30 to 60 minutes. Cast your agarose gel while the tubes are incubating.

37°C is the optimal growth temperature for the bacteria from which these enzymes were isolated.

PROTOCOL 2 PREPARING 1× TBE AND THE AGAROSE GEL

A. Prepare 1× TBE Running Buffer

TBE is usually made as a 10× stock solution.

TBE is used as the electrophoresis running buffer and to prepare the agarose gel.

1. Prepare 350 ml of 1× TBE buffer for your gel electrophoresis. Determine the volume of 10× TBE stock required; document calculations below.

2. Using a 50-ml graduated cylinder, measure the appropriate volume of 10× TBE and pour it into a 500-ml graduated cylinder.

Do not use tap water: minerals will interfere with electrophoresis.

3. Dilute the 10× TBE with distilled or deionized water to a final volume of 350 ml.

B. Preparing the Agarose Gel

0.8% agarose provides the optimal pore size for resolution of lambda DNA.

1. Calculate the amount of agarose needed to prepare 50 ml of 0.8% concentration. Document calculations below.

2. Pour 50 ml of 1× TBE into a 125-ml Erlenmeyer flask. Add the powdered agarose.

3. Microwave the agarose solution at the highest power for about 1 minute.

While the agarose solution is cooling, begin protocol 3.

4. When the agarose solution is cool enough that the flask can easily be held in your hand, add 2 drops of Carolina Blu concentrate and gently swirl to mix. The agarose is ready to be poured.

PROTOCOL 3 CASTING THE AGAROSE GEL AND PREPARING THE GEL CHAMBER

A. Cast the Gel

1. Seal the ends of the casting tray and insert the comb in the notch closest to the end of the tray. Set the casting tray in a place on your lab bench where it will not be disturbed.

Imperfections caused by moving the gel will skew your results.

2. Pour the warm 0.8% agarose solution, containing Carolina Blu, into the casting tray to a depth of about 5 mm. The gel will appear cloudy when it has solidified. Do not move the casting tray during solidification.

B. Prepare the Gel and Gel Chamber

The negative pole is usually color-coded black; the positive is red.

1. When the agarose gel has solidified, unseal the ends of the casting tray and place it in its correct position in the gel chamber. Make sure the comb is at the negative pole.

TBE is used to conduct the current through the gel.

2. Fill the chamber with just enough 1× TBE buffer to cover the entire surface of the gel. Add 10 drops of Carolina Blu final stain to the running buffer in the gel chamber.

3. *Gently* remove the comb, taking care not to tear the gel. Make sure the wells left by the comb are completely submerged. Add more buffer if necessary. The gel is now ready to be loaded with the DNA digests.

PROTOCOL 4 LOADING THE GEL AND PERFORMING GEL ELECTROPHORESIS

EXERCISE 7

1. Remove tubes from the water bath and add 1 µl of loading dye to each tube. Use a fresh tip for each tube. Mix the loading dye with the tube contents by pulsing the tube in a centrifuge.

Loading dye makes the digest denser than the buffer; the digest will sink to the bottom of the well.

2. Obtain an aliquot of DNA marker containing loading dye.

Loading dye provides a means to visually assess the progress of electrophoresis.

3. Pipette 10 µl from each tube, including the DNA marker, into separate wells in the gel chamber (Fig. 7.4). Use a fresh tip for each tube. Correlate and record sample identification with well number.

4. Slide the cover of the chamber into place and connect the electrical leads to the power supply. Make sure the leads are connected so that the DNA migrates in the correct direction.

A stream of fine bubbles at the positive electrode (red) is O_2 gas and at the negative electrode (black) is H_2 gas, indicating current flow.

5. Turn on the power supply. Electrophorese at 120 to 150 volts until the purple dye front is near the end of the gel.

Shortly after the current is applied, loading dye can be seen moving through the gel toward the positive electrode.

6. When electrophoresis is complete, after approximately 40 to 60 minutes, **turn off the power supply.** Disconnect the leads from the power supply and remove the top of the gel chamber.

Fragments of DNA of approximately 2,800 bp migrate with the bromophenol blue (blue dye).

7. Remove the casting tray from the gel chamber. *The gel is very slippery. Hold the tray level.* Slide the gel, right side up, into a staining tray.

Fragments of DNA of approximately 250 bp migrate with the xylene cyanol (purple dye).

8. Discard the buffer by pouring it into the sink.

PROTOCOL 5 STAINING AND DESTAINING COMPLETED GELS

1. Flood the gel with Carolina Blu final stain and incubate for 15 minutes. Wear protective lab clothing and gloves. The stain will discolor your clothes and hands.

DNA is not naturally pigmented; thus it is not immediately visible in a gel.

2. Pour the Carolina Blu final stain back into the container using a funnel. **Work over a sink, with your partner, wearing gloves!**

The stain may be used again; it is recyclable.

3. Retain the gel in the staining tray and gently rinse it with water. Then cover the gel with water and incubate for about 15 minutes.

4. Visualize your DNA bands by placing your gel tray over a white background or by placing the gel on a white light transilluminator. DNA bands appear purplish-blue against a light-blue background.

Bands stained with Carolina Blu intensify if gels are stored in plastic wrap overnight in the refrigerator.

5. Measure the distance each band has migrated, beginning with your size marker lane.

If a camera is available, you may photograph your gel to provide permanent documentation of your results.

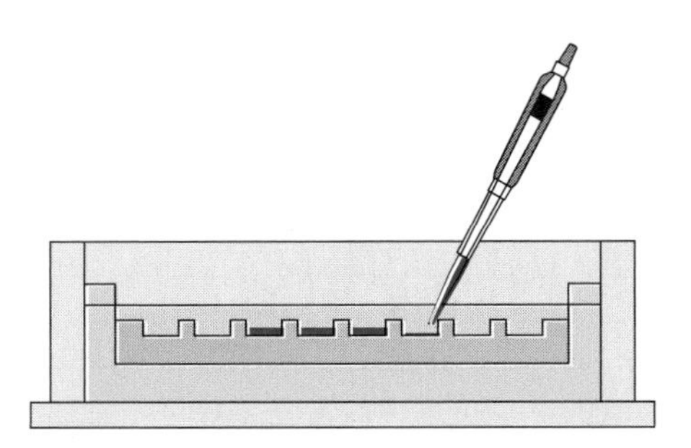

Figure 7.4 Schematic representation of an agarose gel electrophoresis chamber. Wells are being loaded with a micropipette.

6. Record the size marker data in Table 7.1, protocol 6, below.

7. Record your experimental results in Table 7.2, protocol 6, below.

PROTOCOL 6 PREPARATION OF STANDARD CURVE AND ANALYSIS OF RESULTS

1. Use the data from your DNA size marker lane to generate a standard curve. Your instructor will provide you with the number of bands expected in your size marker and the size of each in base pairs (bp).

2. Measure the distance each DNA band has traveled in the gel, in millimeters (mm). Use the front edge of the well as the starting point.

Table 7.1 Standard curve data

Size marker	*x* axis distance migrated (mm)	*y* axis size (bp)
Band 1		
Band 2		
Band 3		
Band 4		
Band 5		
Band 6		
Band 7		

3. Draw your standard curve on semilogarithmic (semilog) graph paper (Fig. 7.5). The *x* axis is linear scale, and the *y* axis is log scale.

4. Draw the best straight line between the points to construct your standard curve.

5. Measure the distance that each DNA fragment has traveled. Record the data in Table 7.2.

6. From your standard curve, extrapolate the size of your DNA fragments and complete Table 7.2.

Table 7.2 Restriction enzyme analysis: experimental data

Band	Lane 1 distance/ size	Lane 2 distance/ size	Lane 3 distance/ size	Lane 4 distance/ size	Lane 5 distance/ size	Lane 6 distance/ size	Lane 7 distance/ size
1							
2							
3							
4							
5							
6							
7							
8							
9							
10							
11							
12							
13							
14							
15							
16							
17							

Materials

Alkaline Lysis Miniprep

Class equipment and supplies

Luria-Bertani (LB) medium
10-mg/ml ampicillin
37°C shaking water bath or incubator
microcentrifuge
disinfectant solution
1 M glucose
1 M Tris, pH 8.0
1 M NaOH
10% sodium dodecyl sulfate (SDS)
1 M Tris, pH 7.4
500 mM EDTA, pH 8.0
3 M sodium acetate (NaOAc), pH 4.8
ice-cold 95% ethanol
ice-cold 70% ethanol
sterile dH_2O
20-mg/ml RNase

Students will need to prepare

LB + ampicillin
overnight culture of bacteria + plasmid
glucose-Tris-EDTA (GTE) solution
NaOH-SDS (NS) solution
10 ml of Tris-EDTA (TE)

Materials per pair of students

starter culture of bacteria + plasmid
Bunsen burner
inoculating loop
sterile microcentrifuge tubes
1 aliquot of 3 M NaOAc, pH 4.8
0.5- to 10-µl micropipette and sterile tips
10- to 100-µl micropipette and sterile tips
100- to 1,000-µl micropipette and sterile tips

Gel Electrophoresis

Class equipment and supplies

microcentrifuge
agarose
10× Tris-borate-EDTA (TBE) buffer (or from exercise 3)
masking tape
dH_2O
DNA marker
ethidium bromide solution
UV transilluminator
Polaroid camera and film

Students will need to prepare

350 ml of 1× TBE
50 ml of 1.0% agarose

Materials per pair of students

gel electrophoresis chamber, gel casting tray, comb
power supply (two pairs of students will share this)
0.5- to 10-µl micropipette and sterile tips
1 aliquot of DNA marker
1 aliquot of loading dye
sterile microcentrifuge tubes
microcentrifuge tube rack
disposable exam gloves
permanent marking pen

Protocols

PROTOCOL 1 INOCULATION OF OVERNIGHT BACTERIAL CULTURE

1. Obtain an agar petri dish with a starter culture of bacteria containing plasmid.

Always work in a sterile manner with bacterial cultures.

2. Prepare 5 ml of LB medium containing 100 µg of ampicillin per ml using sterile LB medium and a 10-mg/ml stock of ampicillin. Document calculations below.

Medium must be cool before antibiotics are added.

3. Flame-sterilize an inoculating loop and pick an isolated colony of bacteria.

Don't forget to cool the inoculating loop before picking a colony.

4. Inoculate the medium by stirring the loop containing bacteria into the LB + ampicillin. Replace the cap of the tube.

5. Incubate the bacterial culture overnight (18 to 24 hours) at 37°C with shaking.

PROTOCOL 2 PREPARATION OF MINIPREP SOLUTIONS

1. Prepare 1 ml of GTE solution. Use the table below to assist with calculations.

Reagent	Stock solution	Final concentration	Volume to make 1 ml
Glucose	1 M	50 mM	
Tris, pH 8.0	1 M	25 mM	
EDTA	500 mM	10 mM	
Volume of dH_2O			
Total volume			1 ml

Measure the dH_2O first and add the glucose, Tris, and EDTA to the dH_2O. Mix by flicking the tube.

*NS solution **must** be prepared fresh before each use, the day it is used.*

2. Prepare 1 ml of NS solution. Use the table below to assist with calculations.

Reagent	Stock solution	Final concentration	Volume to make 1 ml
NaOH	1 M	0.2 M	
SDS	10%	1%	
Volume of dH_2O			
Total volume			1 ml

Measure the dH_2O first and add the NaOH and SDS to the dH_2O. Mix by flicking the tube.

EDTA chelates Mg^{2+}, which is required for DNase functioning.

3. Prepare 1 ml of TE. Use the table below to assist with calculations.

Reagent	Stock solution	Final concentration	Volume to make 1 ml
Tris, pH 7.4	1 M	10 mM	
EDTA, pH 8	500 mM	1 mM	
Volume of dH_2O			
Total volume			1 ml

Measure the dH_2O first and add the Tris and EDTA to it. Add 10 µl of 20-mg/ml RNase to the TE.

PROTOCOL 3 MINIPREP OF PLASMID DNA BY ALKALINE LYSIS

1. Pour about 1.5 ml of the overnight bacterial culture into a microcentrifuge tube.

2. Pellet the bacterial cells by microcentrifuging the culture at 14,000 rpm for 1 minute.

3. Discard the supernatant into the disinfectant solution.

4. Add 100 µl of GTE solution to the pelleted bacteria and resuspend by pipetting up and down.

5. Add 200 µl of the *freshly prepared* NS solution and mix by inverting the tube several times

NS solution denatures linear, chromosomal DNA.

SDS complexes with proteins.

6. Incubate the tube on ice for 1 to 3 minutes.

Do not overincubate.

7. Add 200 µl of 3 M NaOAc, pH 4.8, to the tube. A white precipitate should form.

High salt concentration and neutralization of pH cause linear DNA to reassociate and protein-SDS complexes to precipitate.

8. Incubate the tube on ice for at least 30 minutes.

Tubes may be incubated overnight at 4°C.

9. Pellet the precipitate by microcentrifuging the tube at 14,000 rpm for 10 minutes.

10. Remove 400 µl of supernatant into a clean microcentrifuge tube.

11. Add 1 ml of ice-cold 95% ethanol to the tube containing the supernatant. Mix by inverting the tube.

12. Precipitate the DNA by incubating the tube on ice for 30 minutes.

DNA may be precipitated overnight at −20°C.

13. Pellet the DNA by microcentrifuging at 14,000 rpm for 10 minutes.

14. Carefully pour off the ethanol.

15. Wash the DNA by filling the tube with ice-cold 70% ethanol and pouring it out. Take care not to dislodge the DNA pellet.

Washing removes contaminating salt.

16. Air dry the DNA pellet to allow the ethanol to evaporate. This may require overnight incubation.

Ethanol contamination will interfere with subsequent analysis.

first try 20ml, then 40ml.

17. Add 40 µl of TE to the pellet and dissolve the DNA into solution.

18. Mix 10 µl of sample with 1 µl of loading dye and analyze the plasmid purification by gel electrophoresis.

This DNA may also be digested with restriction enzymes before gel electrophoretic analysis.

PROTOCOL 4 PREPARING 1× TBE AND THE AGAROSE GEL

A. Prepare 1× TBE Running Buffer

1. Prepare 350 ml of 1× TBE for your gel electrophoresis. Determine the volume of 10× TBE stock required; document calculations below.

TBE is usually made as a 10× stock solution.

TBE is used as the electrophoresis running buffer and to prepare the agarose gel.

2. Using a 50-ml graduated cylinder, measure the appropriate volume of 10× TBE and pour it into a 500-ml graduated cylinder.

3. Dilute the 10× TBE with distilled or deionized water to a final volume of 350 ml.

Do not use tap water: minerals will interfere with electrophoresis.

B. Preparing the Agarose Gel

1. Calculate the amount of agarose needed to prepare 50 ml of 1.0% concentration. Document calculations below.

2. Pour 50 ml of 1× TBE into a 125-ml Erlenmeyer flask. Add the powdered agarose.

3. Microwave the agarose solution at the highest power for about 1 minute.

While the agarose solution is cooling, begin protocol 5.

4. When the agarose solution is cool enough that the flask can easily be held in your hand, the agarose gel is ready to be poured.

PROTOCOL 5 CASTING THE AGAROSE GEL AND PREPARING THE GEL CHAMBER

A. Cast the Gel

1. Seal the ends of the casting tray and insert the comb in the notch closest to the end of the tray. Set the casting tray in a place on your lab bench where it will not be disturbed.

Imperfections caused by moving the gel will skew your results.

2. Pour the warm 1.0% agarose solution into the casting tray to a depth of about 5 mm. The gel will appear cloudy when it has solidified. Do not move the casting tray during solidification.

B. Prepare the Gel and Gel Chamber

The negative pole is usually color-coded black; the positive pole is red.

1. When the agarose gel has solidified, unseal the ends of the casting tray and place it in its correct position in the gel chamber. Make sure the comb is at the negative pole.

TBE is used to conduct the current through the gel.

2. Fill the chamber with just enough 1× TBE to cover the entire surface of the gel.

3. *Gently* remove the comb, taking care not to tear the gel. Make sure the wells left by the comb are completely submerged. Add more buffer if necessary. The gel is now ready to be loaded with the plasmid preparation.

PROTOCOL 6 LOADING THE GEL AND SETTING UP THE GEL CHAMBER FOR ELECTROPHORESIS

Fragments of DNA of approximately 2,800 base pairs migrate with the bromophenol blue (blue dye).

Fragments of DNA of approximately 250 base pairs migrate with the xylene cyanol (purple dye).

1. Carefully pipette the plasmid sample into a well in the agarose gel. Don't forget to include a DNA marker and to use a fresh tip for each sample. Correlate and record sample identification with well number.

2. Slide the cover of the chamber into place and connect the electrical leads to the power supply. Make sure your leads are connected so that the DNA migrates the correct direction in the gel.

A stream of fine bubbles at the positive electrode is O_2 gas and at the negative electrode is H_2 gas, indicating current flow.

3. Turn on the power supply. Electrophorese your gel at 120 to 150 volts. You will run your gel until the purple dye front is near the end of the gel.

Shortly after the current is applied, loading dye can be seen moving through the gel toward the positive electrode.

4. When the electrophoresis is complete, after approximately 40 to 60 minutes, turn off the power supply. Disconnect the leads from the power supply, and remove the top of the gel chamber.

5. Discard the buffer by pouring it down the sink.

PROTOCOL 7 ETHIDIUM BROMIDE STAINING OF DNA IN AGAROSE GELS

DNA is not naturally colored; thus it is not immediately visible in a gel.

Ethidium bromide is a mutagen and suspected carcinogen.

1. **Wearing gloves**, remove the casting tray from the gel chamber. *The gel is very slippery. Hold the tray level.* Slide the gel, right side up, into a 1-µg/ml ethidium bromide solution.

2. Stain the gel for 5 to 10 minutes at room temperature.

3. Visualize the DNA bands by placing your gel on a UV light transilluminator. The ethidium bromide stain is fluorescent.

4. Turn the transilluminator on and photograph the gel with a Polaroid camera and black-and-white film. Use an orange filter. Approximate settings are an f-stop of 8 with a 1/2-second exposure.

Fluorescence intensity is proportional to the amount of DNA present.

UV light can cause skin and eye burns.

References and General Reading

Ausubel, F. M., R. Brent, R. E. Kingston, D. D. Moore, J. G. Seidman, J. A. Smith, and K. Struhl (ed.). 1994. *Current Protocols in Molecular Biology.* Wiley Interscience, New York, N.Y.

Birnboim, H. C. 1983. A rapid alkaline extraction method for the isolation of plasmid DNA. *Methods Enzymol.* **100:**243–255.

Discussion

1. Discuss the components and function of each solution used in the miniprep protocol.
2. Examine the photo of the agarose gel results of your miniprep sample. How many bands do you observe? What type of DNA does each band represent?
3. Discuss at least two methods for determining the quantity of plasmid DNA isolated. Which would you use and why?
4. Discuss potential uses for plasmid DNA isolated using a miniprep protocol.

EXERCISE 9

Bacterial Transformation

Key Concepts

1. Bacterial transformation is a process in which naked fragments of DNA are taken up into a bacterial cell. This process occurs naturally, or bacteria can be made competent to accept foreign pieces of DNA in the laboratory.
2. Bacterial transformation is important in recombinant DNA technology because transformed bacteria can be utilized to clone foreign DNA inserted into a plasmid.
3. Bacteria can be induced to translate proteins from the DNA contained on a plasmid. One example is the production of human insulin to treat diabetes.

Goals

By the end of this exercise, you will be able to

- discuss bacterial transformation
- understand gene cloning
- discuss recombinant DNA
- understand how recombinant DNA technology can be used

Introduction

In 1928 Fred Griffith first described the process of transformation. He showed that living nonpathogenic bacteria could acquire characteristics from dead disease-causing bacteria and become transformed into disease-producing ones. Oswald Avery, MacLyn McCarty, and Colin MacLeod showed that Griffith's transforming principle was, in fact, DNA. Scientists later discovered that bacteria in nature can accept pieces of naked DNA from their environment. These naked fragments of DNA may come from dead bacteria of the same or a different species. The process wherein bacteria take up fragments of DNA is called transformation.

When transformation occurs, the DNA transferred is often a plasmid: small, circular DNA found naturally in many bacteria (Fig. 9.1). A plasmid exists separately from the bacterium's own chromosomal DNA and contains genes that the bacterium would not normally possess. These extra genes can provide a growth advantage to bacteria by providing the gene for an enzyme such as amylase, which breaks down starches, or beta-lactamase, which breaks down the antibiotic ampicillin. Thus, the bacteria can grow where they normally might not.

Plasmids in nature vary greatly in size; some are only a few thousand base pairs, and others are over 200 kilobase pairs (kbp). Most plasmids used in DNA technology are only 4 to 5 kbp in size, whereas the *Escherichia coli* chromosome is about 4,000 kbp. Every time the bacterium replicates its chromosomal DNA, it replicates the plasmid, too. Some plasmids are not under such stringent control and may replicate independently as well.

In nature, bacteria that are naturally competent may acquire plasmids containing antibiotic resistance genes. This is critical to the field of medicine. New strains of bacteria resistant to common antibiotics are developing in hospitals.

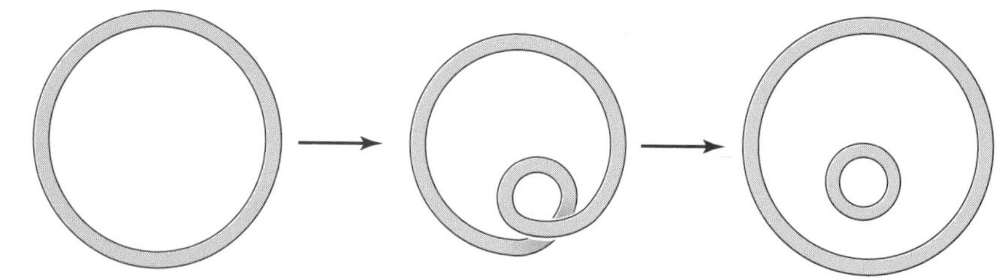

Figure 9.1 Plasmids probably evolved early in the evolution of living organisms by this mechanism. An internal loop forms from the normal bacterial chromosome. The ends of the loop join each other and separate from the main chromosome, forming a plasmid.

Plasmid-encoded resistance genes can spread through antibiotic-susceptible bacterial populations. New antibiotics, which are very difficult to develop, must be found to fight infections once thought to be under control.

Of great interest to molecular biologists, however, is the fact that plasmids can be genetically engineered. Any gene can be isolated from human DNA and inserted into a plasmid. This recombinant plasmid, containing a new human gene, can be transformed into a suitable bacterial host. This process allows the researcher, by culturing the transformed bacteria, to make large quantities of the gene of interest. Next, researchers can begin to characterize a new gene. The first step is to sequence the inserted gene, that is, to determine the exact order of nucleotides that make up the gene of interest.

Molecular biologists have developed techniques to make transformed bacteria produce proteins from genes inserted into plasmids. These proteins can then be purified from bacteria and used for many purposes. One example is the human insulin gene. The DNA sequence coding for insulin is inserted into a special type of plasmid called an expression vector. Within this plasmid is a promoter sequence to initiate transcription of the foreign DNA. The plasmid expression vector is then transformed into a bacterial host. Translation of the messenger RNA thus produced yields large quantities of inactive insulin. Posttranslational modification follows; then the active insulin is purified for use in diabetes treatment.

pGFPuv

Two elements are required in a transformation system. The first element is a suitable host bacterium. This exercise uses *E. coli*, a bacterium commonly found in our intestines. The strain of *E. coli* used here has been cultured in the laboratory for so many years that it is not likely to be capable of colonizing the human intestine. Also, it has been selected for characteristics that make it especially useful in the molecular biology laboratory. Because of these special properties, however, the laboratory strain does not typically grow in the environment.

The other important element in the transformation system is the plasmid. The plasmid we will use is called pGFPuv (Fig. 9.2). Plasmid pGFPuv contains a gene that encodes green fluorescent protein (GFP) and the ampicillin resistance gene, which encodes beta-lactamase, an enzyme that breaks down the antibiotic ampicillin. Expression of the genes for GFP and beta-lactamase will provide us with a mechanism for determining whether the transformation experiment has been successful.

GFP is a 238-amino-acid, 28-kDa protein found in *Aequorea victoria*, a jellyfish. The production of the protein causes the jellyfish to fluoresce and glow in the dark. Scientists isolated the wild-type gene (wt-GFP) from *A. victoria* and modified it by changing the DNA coding for three amino acids: Phe-99 to Ser, Met-153 to Thr, and Val-163 to Ala. None of these mutations alter the chromophore (fluorescent moiety) sequence but instead make the GFPuv protein fluoresce 18 times brighter than wt-GFP. The modified gene was then placed into a plasmid that was subsequently named pGFPuv.

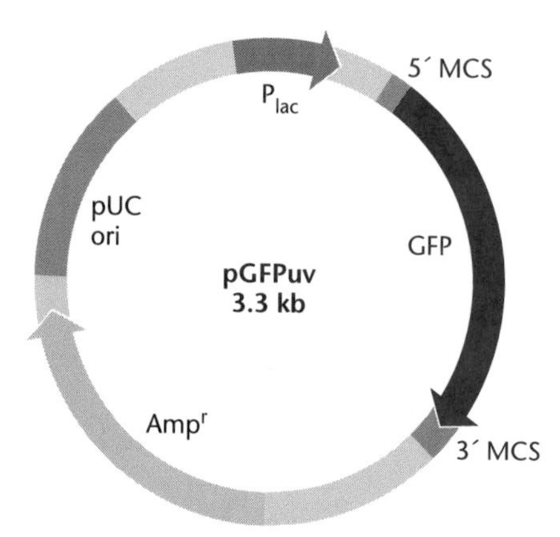

Figure 9.2 Plasmid pGFPuv, which contains the genes for GFP and beta-lactamase. MCS, multiple cloning site.

When *E. coli* is transformed with pGFPuv, the bacteria express the modified GFPuv. The fluorescence provided by the GFPuv is detected by exposure to standard long-wave ultraviolet (UV) light, either from a UV transilluminator or a handheld UV lamp. The GFP expressed in *E. coli* is soluble and of low toxicity; therefore, the *E. coli* containing GFPuv grow normally while expressing the protein efficiently and for a long time.

pGFPuv can also be used as a bacterial cloning vector. The plasmid vector contains a *lacZ* promoter that expresses GFPuv when induced with isopropyl-β-D-thiogalactopyranoside (IPTG). In addition, it has a multiple cloning site and contains the pUC origin of replication (ori) and an ampicillin resistance gene for propagation and selection in *E. coli.*

Transformation in the Laboratory

Although some bacteria can accept plasmids naturally from other bacteria in the environment and express proteins from plasmid genes, *E. coli* must be made competent to take up foreign DNA in a laboratory setting. Bacterial cells are made competent by incubation on ice in calcium chloride. This apparently makes the cell envelope more permeable. The next step is a heat shock step. Bacterial cells are moved from 0°C (ice) to 42°C and then back to ice again. This supposedly alters the cell envelope, allowing the plasmid to slip inside, then closes the cell envelope again. The final step of the transformation process is a recovery step. A short period of incubation in Luria-Bertani (LB) broth, a nutritionally complete liquid growth medium, enables the cells to recover from the experimental manipulations.

Once the transformation process has been completed, the experiment must be assayed by growing the bacteria under a variety of conditions to determine if transformation occurred. LB agar plates and LB agar plates incorporating ampicillin are available. You must design an assay to determine if transformation occurred.

Electroporation

The method used in this laboratory exercise to make bacteria competent and to transform them with plasmid DNA is very quick and easy, but not very efficient. The bacteria are not quantitated in any way, nor are they treated with optimal conditions for permeabilizing the cell membrane. Furthermore, various strains of bacteria may transform more efficiently under various conditions. When high efficiency (transformants per microgram of plasmid DNA) is desired, electroporation is the method of choice. Electroporation uses a high-voltage charge to make transient holes in the bacterial membrane through which the plasmid may move. At present, it is considered to be the most efficient method for transforming bacterial cells. The transformation efficiency of *E. coli* that is transformed by using $CaCl_2$ ranges from 10^6 to 10^8 transformants per microgram of plasmid DNA. A researcher using electroporation may expect 5×10^9 to as high as 1×10^{11} transformants per microgram of DNA.

Lab Language

aliquot
competent
control
efficiency
electroporation
incubate/incubation
negative control
plasmid
positive control
procedure/protocol
solution
transformants
transformation
vortex

Laboratory Safety

Review the sections on physical, biological, and chemical hazards in "Laboratory Safety." Special considerations for this exercise:

working with microbes
UV light source

Materials

Class equipment and supplies
42°C water bath
37°C incubator
UV light source

Materials per pair of students
1,000-ml beaker to collect biological waste
wax marking pencil
1 plate of wild-type *E. coli* (strain DH5α) grown on LB agar
container with ice
2 microcentrifuge transformation tubes
1 aliquot of 50 mM $CaCl_2$
1 aliquot of plasmid, 0.005 μg/μl (Clontech)
1 aliquot of 1× Tris-EDTA
inoculating loop
Bunsen burner
2 LB agar plates
2 LB agar + ampicillin plates (100-μg/ml ampicillin)
0.1- to 10.0-μl micropipette and sterile tips
100- to 1,000-μl micropipette and sterile tips
1 aliquot of sterile LB broth
500-ml beaker containing ethanol
spreading rods

Protocols

PROTOCOL 1 PREPARATION OF COMPETENT BACTERIAL CELLS AND TRANSFORMATION MIXTURES

Keep tubes on ice.

Cold $CaCl_2$ is important for bacterial cell competence.

1. Add 250 μl of 50 mM $CaCl_2$ to two tubes. Place the two tubes on ice. Label test tubes with group number or initials.

2. Use a sterile inoculating loop to aseptically transfer a single colony of bacteria from the starter plate to each tube. Mix vigorously with the inoculating loop and return to ice.

Proper resuspension of bacterial cells ensures exposure to plasmids in solution.

3. Immediately place the two tubes into the vortex shaker for a few seconds to break up the clumps of bacterial cells. (Carry the tubes on ice to the vortex.) If a vortex is not available, flick the tubes vigorously with your finger. All the clumps of bacteria should be broken up.

The suspension with plasmid is the experimental transformation.

4. Add 10 μl of the plasmid (0.005-μg/μl) solution directly into the cell suspension in *one* of your tubes. Label the tube. Tap the tube directly with your finger to mix. Avoid making bubbles.

5. Return this tube to ice.

The suspension with 1× Tris-EDTA is the control transformation.

6. Add 10 μl of 1× Tris-EDTA to the second tube of bacterial cell suspension. Label the tube. Tap the tube directly with your finger to mix. Avoid making bubbles.

7. Return this tube to ice.

8. Incubate both tubes on ice for 15 minutes.

PROTOCOL 2 DECIDING HOW TO DETECT TRANSFORMATION

To see if you successfully transformed your cells, you will need a plan to make that determination. LB agar plates and LB agar plates with the ampicillin antibiotic added are available for you to use. You will need four plates:

- an experimental plate
- a negative control plate
- two positive control plates

Use the following diagrams, representing agar plates, to plan the plating of your experiment. Label each plate with the type of medium it will contain and the type of bacteria you will spread on each. Include your name and the date on each plate. Discuss your plan with an instructor before proceeding.

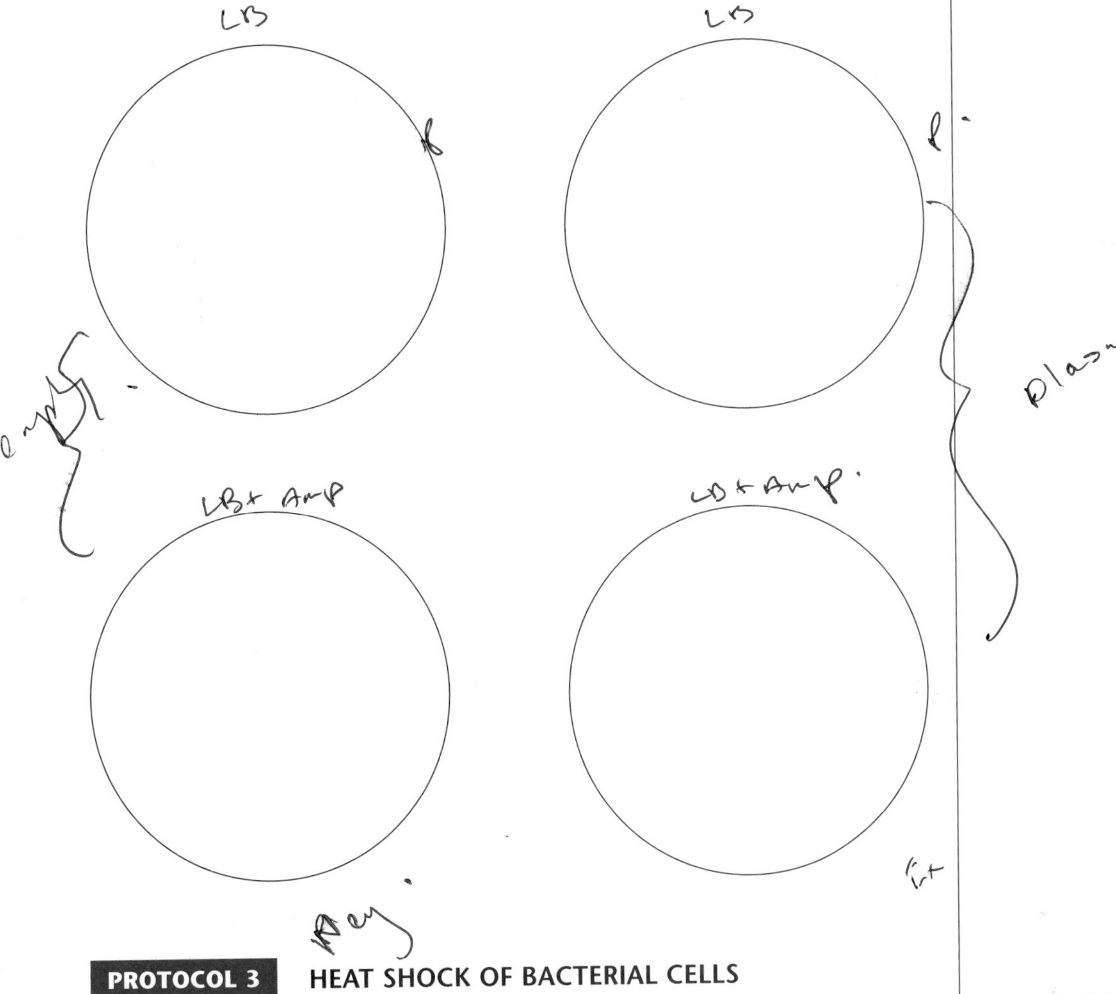

PROTOCOL 3 HEAT SHOCK OF BACTERIAL CELLS

Heat shock helps the cells to take up the plasmids.

Timing is critical.

1. Carry your tubes on ice to the water bath. Remove the tubes from ice and *immediately* immerse them in the 42°C water bath (or heat block) for 90 seconds. It is *critical* that cells receive a sharp and distinct shock.
2. *Immediately* return both tubes to ice and let them stand on ice for 1 additional minute.
3. Place the tubes in a test tube rack at room temperature.
4. Add 250 µl of sterile liquid LB medium to each tube. Gently tap the tubes to mix.

5. Allow the bacteria to recover from the heat shock at room temperature for 30 to 45 minutes.

During recovery, bacterial cells repair their membranes, grow, and express the genes acquired with the new plasmid DNA.

PROTOCOL 4 PLATING THE TRANSFORMATION

1. Using the plan you devised in protocol 2 to detect transformation, pipette 100 µl of cells onto each plate.

Use proper aseptic technique when inoculating the plates.

2. Using a sterile cell spreader, spread the bacteria over the surface of the agar in each plate. Move the spreader back and forth on the agar while turning the plate. This will spread the bacteria evenly across the agar surface.

See exercise 4.

3. Place the plates upside down in a 37°C incubator and allow them to grow for 24 to 48 hours. Alternatively, plates may incubate at room temperature for 2 to 3 days.

Inverting the plates prevents condensation from collecting on the medium.

4. When colonies are visible, place the plates on a UV transilluminator or illuminate with a handheld UV source.

Cultures observed after 72 hours will produce only a faint glow.

5. Observe and count the number of glowing colonies.

References and General Reading

Ausubel, F. M., R. Brent, R. E. Kingston, D. D. Moore, J. G. Seidman, J. A. Smith, and K. Struhl (ed.). 1994. *Current Protocols in Molecular Biology.* Wiley Interscience, New York, N.Y.

Bloom, M. V., G. A. Freyer, and D. A. Micklos. 1996. *Laboratory DNA Science.* Benjamin/Cummings, New York, N.Y.

Chalfie, M., Y. Tu, G. Euskirchen, W. W. Ward, and D. C. Prasher. 1994. Green fluorescent protein as a marker for gene expression. *Science* **263:**802–805.

Cormack, B. P., R. H. Valdivia, and S. Falkow. 1996. FACS-optimized mutants of the green fluorescent protein (GFP). *Gene* **173:**33–38.

Cubitt, A. B., R. Heim, S. R. Adams, A. E. Boyd, L. A. Gross, and R. Y. Tsien. 1995. Understanding, improving and using green fluorescent proteins. *Trends Biochem. Sci.* **20:**448–455.

Hanahan, D. 1983. Studies on transformation of *Escherichia coli* with plasmids. *J. Mol. Biol.* **166:**557–563.

Heim, R., A. B. Cubitt, and R. Y. Tsien. 1995. Improved green fluorescence. *Nature* **373:**663–664.

Prasher, D. C., V. K. Eckenrode, W. W. Ward, F. G. Prendergast, and M. J. Cormier. 1992. Primary structure of the *Aequorea victoria* green-fluorescent protein. *Gene* **111:**229–233.

Discussion

1. Describe the process that makes bacterial cells competent.
2. Why do bacterial cells need to be competent for a successful transformation?
3. How do researchers use plasmids?
4. Use the following diagrams to draw your results. Label the type of medium in each plate and the type of bacteria you spread on each. Count the number of colonies, where applicable.

5. Describe the growth on your positive control plates. Is this what you expected to observe? Explain.
6. Describe the growth on your negative control plate. Is this what you expected to observe? Explain.
7. Describe the growth on your experimental plate. Is this what you expected to observe? Explain.
8. Was your transformation experiment successful? Explain.
9. Did you detect any contamination on your plates? How would you know that it was contamination?
10. Can you count individual colonies on your experimental plate? If so, how many do you have?

Transformation efficiency (TE) can be determined if you know how much plasmid DNA you started with. In this experiment, you started with 0.005 μg/μl of pGFPuv. TE = number of transformants per microgram of DNA in plated bacteria.

11. What was the *total* mass of plasmid DNA used at the beginning of your experiment?
12. What is the mass of plasmid DNA in the *plated* bacteria? (Show your work.) Hint:

$$\mu\text{g of DNA in the plated bacteria} = \frac{\text{total mass of DNA used}}{\text{total volume}} \times \text{volume of bacteria plated}$$

where total volume = volume of $CaCl_2$ + volume of LB medium.

13. What is the transformation efficiency? (Show your work.)

14. What does the transformation efficiency tell you about the result of your experiment?
15. When you examine your plates after a few days of incubation, you may see tiny colonies surrounding a larger colony. These are called satellite colonies and are bacteria that probably did not successfully transform with the plasmid. Design an experiment to prove this hypothesis. Be sure to include a control.
16. How were the satellite colonies able to grow on a plate with ampicillin if they did not take up a plasmid?
17. How is a recombinant plasmid made?
18. It is very useful to have the gene for ampicillin resistance present on a plasmid when you clone a new gene into that plasmid. Can you think of a reason why?

EXERCISE 10

Amplification of Human DNA by the Polymerase Chain Reaction

Key Concepts

1. Polymerase chain reaction (PCR) is a very powerful technique for making billions of copies of small segments of DNA.
2. PCR can be performed successfully with minute quantities of DNA. Also, PCR can amplify DNA that has been degraded.
3. PCR technology can be used to diagnose diseases, find gene mutations, and identify individuals.

Goals

By the end of this exercise, you will be able to

- discuss PCR
- understand DNA polymorphisms
- discuss the TPA-25, APOC2, and D1S80 loci and their usefulness in studies of human DNA
- describe the use of oligonucleotide primers in PCR
- understand the applications of PCR in biological and medical science

Introduction

During the mid-1980s, Kary Mullis developed an elegant and powerful technique to generate millions of copies of a specific DNA sequence. This technique is called PCR. It amplifies, or copies, the desired DNA. The term amplification is used because scientists can start with a quantity of DNA undetectable by standard methods. The process of PCR produces target DNA in quantities sufficient for detection by agarose gel electrophoresis. Currently, PCR can most efficiently amplify fragments of DNA of 1,000 base pairs (bp) or less, but technology is improving to provide amplification of longer pieces.

PCR is analogous to photocopying: we start with an original (the desired DNA sequence), use the appropriate machinery (a thermal cycler) and reagents (DNA polymerase, nucleoside triphosphates, primers), and churn out millions of identical DNA sequences, just as a photocopy machine churns out many copies of an original document.

The primers are a critical component in PCR, providing specificity for the desired region. Two oligonucleotide primers are selected to flank the DNA sequence of interest. A primer is a short, single-stranded sequence of DNA, an oligonucleotide, capable of binding to a region of DNA adjacent to the desired target sequence. Two different primers are used: one binds to one strand at the beginning or 5′ end of the desired DNA, the other to the opposite strand at the 3′ end of the desired sequence (Fig. 10.1). The primers isolate the specific region of DNA to be amplified. Each primer functions by binding specifically to its comple-

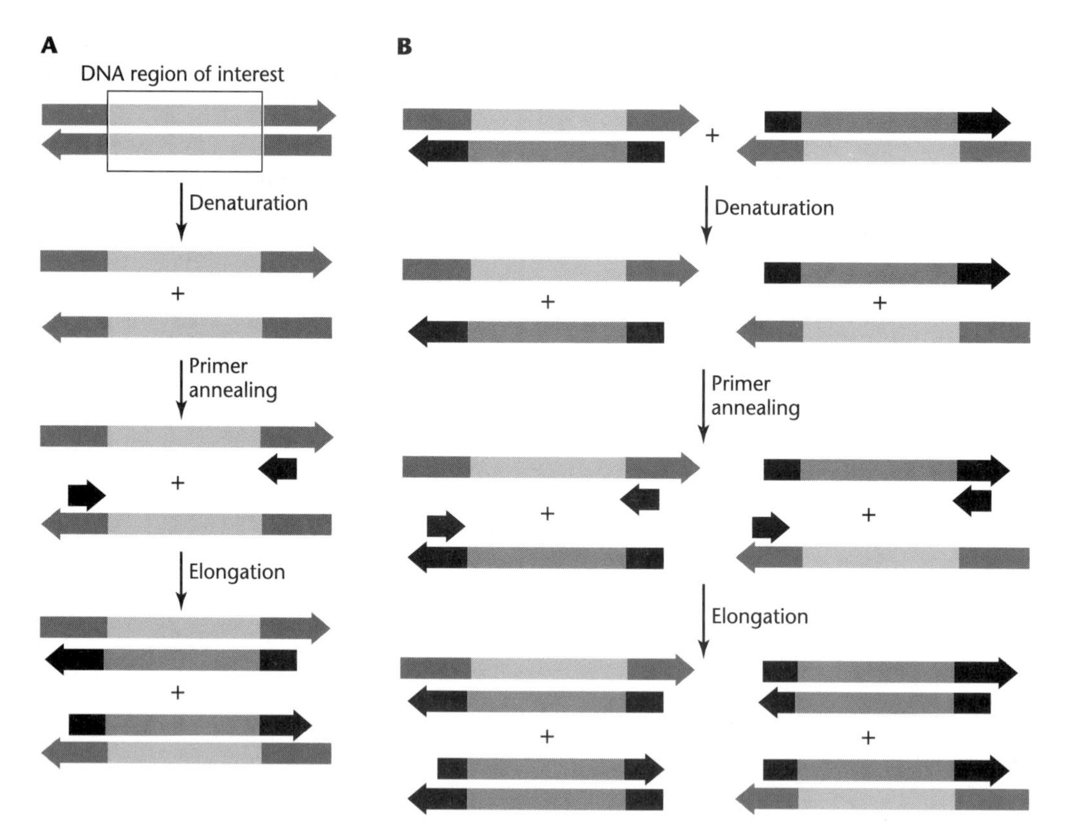

Figure 10.1 PCR. (A) One cycle of PCR doubles the template DNA. (B) The second and subsequent PCR cycles use the newly synthesized DNA as template DNA. DNA copies are expanded exponentially.

mentary stretch of DNA on the template. Once the primers have bound to the respective ends of the desired DNA, the enzyme DNA polymerase copies the target DNA template using the bound primers as a starting point. In this way, both strands of target DNA are replicated.

Template DNA must be included in the reaction. DNA can be obtained from many sources, such as cheek cells, a hair follicle, a fingernail, or blood. A few cells or bits of tissues will supply enough DNA for the reaction. The DNA may also be degraded by exposure to the elements, as at a crime scene or an archeological dig.

All four nucleoside triphosphates, adenine (A), cytosine (C), guanine (G), and thymine (T), are required. These are the building blocks of DNA.

Buffer is required to provide the correct environment for the DNA polymerase enzyme to function. Buffer maintains pH and provides the correct concentration of magnesium, without which the enzyme will not function. Varying the concentration of the magnesium affects the efficiency of PCR.

PCR requires DNA polymerase. It replicates the target DNA by catalyzing the reaction between the 5′ phosphate group of one nucleotide and the 3′ hydroxyl of the growing DNA strand. *Taq* DNA polymerase, isolated from *Thermus aquaticus*, a thermophilic bacterium inhabiting volcanic hot springs, does not denature as rapidly as other proteins at high temperatures. Thermal stability of *Taq* polymerase is essential in the temperature cycling required for PCR.

Temperature cycling is essential to PCR. After the template DNA and all reagents have been mixed, the cycling process of PCR begins. There are three steps to each cycle of PCR: denaturation, annealing, and elongation (Fig. 10.1), each of which requires a specific temperature.

Denaturation is the first step and occurs by heating the reaction mixture to 94°C. This step breaks the hydrogen bonds that maintain the double-stranded configuration. The template DNA separates into its single-stranded components.

Annealing is the second step of the PCR cycle. The temperature is lowered so that the primers can form hydrogen bonds with complementary bases on the template DNA. Temperatures of 45 to 72°C are frequently used. Temperature is critical. The temperature selected depends on the length and sequence of the primer and, therefore, the number of hydrogen bonds that must be formed. Primer binding to DNA at higher temperature conditions, those that would favor DNA strand separation, is more difficult to achieve. This condition is referred to as stringent. Decreasing the stringency by lowering the temperature allows the primers to bind with some errors so that a homologous, but not always identical, gene can be isolated. Very stringent conditions would be used in forensic applications.

Elongation completes the PCR cycle. The temperature is raised to 72°C, not hot enough to denature the primer tightly bound to its complement, but hot enough to denature the primer molecules loosely bound with mismatches and thereby prevent any additional nonspecific priming. The *Taq* DNA polymerase then catalyzes the addition of nucleotides to the primer, elongating the primer by reading the complementary template DNA strand.

One cycle of PCR is now completed, giving us two copies of the desired DNA. Once a new DNA strand has been made via the first PCR cycle, it serves as a template molecule in subsequent cycles. After each cycle of DNA denaturation, primer annealing, and elongation of the complementary strand, the number of molecules of desired DNA doubles. For example, after two cycles, we have four target DNA molecules; after four cycles, 16 target molecules; and after a typical PCR of 30 cycles, taking only a few hours in an automated thermal cycler, we have over 1 billion copies of target DNA.

PCR has revolutionized all areas of science. Many researchers have used the technique of PCR, in one way or another, in their work. Applications include forensic analysis, diagnostics, DNA sequencing, mutagenesis, cloning, and gene isolation.

PCR is especially useful when very little DNA is available. DNA from a tiny spot of blood can be amplified and used to incriminate or exonerate a suspect in a criminal investigation. DNA in fossils can be amplified and compared with DNA from living organisms to determine evolutionary relationships. PCR can be used to identify human immunodeficiency virus and other viruses and to diagnose sickle cell anemia as well as other genetic disorders.

PCR is widely used in forensics because of two important features. First, only one copy of a region of DNA is necessary to begin PCR amplification. The process of PCR amplifies DNA logarithmically. One copy of DNA, amplified through 30 cycles of PCR, ends with over 1 billion copies of that region. In this manner, very small samples of blood, semen, hair, or DNA preserved in other sources can be analyzed. The second feature of PCR is that the DNA does not have to maintain its integrity. It doesn't matter if the source DNA is partly degraded, contaminated with other biochemicals, or found within clothing or complex tissue. PCR ignores everything except the desired DNA sequence. DNA degraded by exposure to the elements at a crime scene or broken down by age, as in samples from mass grave sites, still can be amplified.

PCR is very useful in diagnosis of infectious diseases, especially those caused by viruses. Viral infections typically are difficult to diagnose because the growth and culture of viruses are laborious. Antibody testing is inconvenient. Two serum samples, an acute sample taken at the peak of disease and a convalescent sample taken when the patient is better, are usually required and are tested for antibody levels. To conclusively diagnose viruses by a patient's antibody response, the serum samples must reflect a change in antibody titer. By specifically searching for viral gene sequences, produced only in active infections, PCR provides a rapid and sensitive method for virus detection.

Scientists use PCR to search for homologous genes across species. Researchers frequently find important genes in yeast, bacteria, and *Drosophila melanogaster*, because these organisms can be rapidly propagated and their genetic information easily manipulated. PCR can be used to search for the same gene in humans.

The importance of PCR in the advancement of knowledge was recognized in 1993 when Mullis was awarded the Nobel Prize for his discovery.

TPA-25: an *Alu* Insertion in the Tissue Plasminogen Activator Gene on Chromosome 8

In this laboratory experiment, the region of human DNA to be amplified is the *Alu* element TPA-25, inserted into the tissue plasminogen activator gene located on chromosome 8 (Fig. 10.2). The sequence, about 300 bp in length, is located within an intron of the tissue plasminogen activator gene. Although noncoding, the TPA-25 region is inherited according to Mendelian principles. The *Alu* sequence is dimorphic: it is either present or absent. Some individuals may be homozygous for the insertion, that is, they possess the TPA-25 insertion on both maternal and paternal chromosome 8; some are heterozygous; and some will not have the insertion on either chromosome.

The oligonucleotide primers we are using will amplify a 400-bp fragment from the tissue plasminogen activator gene intron if TPA-25 is inserted. If the *Alu* insertion is absent, the fragment will be only 100 bp in length. These differences in allele size are detectable by agarose gel electrophoresis. PCR of homozygotes will produce a pattern with a single band: either a 400-bp fragment if TPA-25 is present or a 100-bp band if TPA-25 is absent (Fig. 10.2). Heterozygotes will have two bands: 100 bp and 400 bp. Under the experimental conditions to be used, homozygotes with the TPA-25 insertion will be distinguishable from homozygotes lacking the insertion.

Alu elements are scattered throughout the human genome and may compose about 5% of the total human DNA. *Alu* sequences are members of a group of intermediate-repeat DNA sequences in mammals called short interspersed elements (SINEs). *Alu* is a mobile element: it may insert into an exon (as documented in one patient with neurofibromatosis) as well as in between genes or into introns such as TPA-25. In fact, *Alu* is the most abundant human SINE. The *Alu* sequence is extremely similar to that of 7S RNA, a small RNA that is part of the signal recognition particle found in the cytoplasm. The signal recognition particle assists in the movement of new polypeptide through the membranes of the endoplasmic reticulum. It is hypothesized that *Alu* sequences may have evolved from the 7S RNA sequence. *Alu* appears to have become widely distributed throughout the human genome through a retrotransposition mechanism that may involve the reverse transcriptase coded for by L1, another mobile element found in humans. This process is thought to be mediated by an RNA splicing mechanism.

Tissue plasminogen activator exon region
Tissue plasminogen activator intron region
TPA-25 insertion sequence

Tissue plasminogen activator gene sequence without *Alu* insertion

Tissue plasminogen activator gene sequence with *Alu* insertion

Figure 10.2 The tissue plasminogen activator gene region showing no insertion (top) resulting in a 100-bp amplification fragment or with a TPA-25 insertion sequence (bottom) giving a 400-bp product. Individuals may be homozygous for either form or heterozygous.

APOC2

The APOC2 locus codes for apolipoprotein C2, which transports cholesterol in the blood. A variable number of tandem repeat (VNTR) sequence (Fig. 10.3), two nucleotides in length, occurs in one of the introns of the gene. The largest allele identified contains 30 repeat sequences. Amplification of this region results in DNA fragments ranging between 100 and 200 bp in length.

The D1S80 Locus

D1S80 is a region of DNA on the short arm of chromosome number 1 that does not code for protein. Noncoding regions of DNA often contain polymorphisms, that is, many different alleles. These DNA regions are very diverse in the human population. Even though D1S80 does not code for protein, it is still inherited according to Mendel's laws. Each individual gets one allele from the mother and one from the father. Everyone has two copies of the D1S80 locus and may be homozygous or heterozygous, depending on the allele sizes inherited from parents. This

Individual 1

TTAGACGG ACGT ACGT ACGT ACGT ACGT ACGT ACGT CGGGTTCCTT

Individual 2

TTAGACGG ACGT ACGT ACGT ACGT ACGT ACGT ACGT ACGT ACGT ACGT ACGT ACGT ACGT ACGT ACGT CGGGTTCCTT

Figure 10.3 An example of a region of DNA containing a four-nucleotide VNTR sequence. Individual 1 has seven repeats, which will result in a 28-bp-long region of DNA. Individual 2 has 15 repeats, which will result in a 60-bp-long region of DNA. Primers for PCR bind to the region of DNA on either side of the VNTR region.

region is commonly used in forensic analysis to include or exclude suspects in criminal cases. D1S80 may also be used to help determine parentage.

D1S80 is a VNTR DNA polymorphism. The basic unit of the region is a sequence of 16 bp. However, this sequence may be repeated from 14 to 40 times on each of the two number 1 chromosomes. An allele having 14 repeats would be 224 bp long; an allele having 40 repeats would be 640 bp long (Fig. 10.3). The VNTR length on one's maternal chromosome may be different from the VNTR length on one's paternal chromosome (Fig. 10.4). One individual may be heterozygous, having two different alleles yielding two different-sized pieces of amplified DNA. Another individual may be homozygous, yielding pieces of DNA that are all the same size. These differences in allele size are detectable by agarose gel electrophoresis. PCR of homozygotes will produce a pattern with a single band. Heterozygotes will have two bands.

Lab Language

allele
Alu
amplify
annealing
buffer
denaturation
diagnosis
DNA polymerase
elongation
forensics
heterozygous
homozygous
nucleoside triphosphate
PCR
polymorphic
primer
SINEs
temperature cycling
template
VNTR

Laboratory Safety

Review the sections on physical, biological, and chemical hazards in "Laboratory Safety." Special considerations for this exercise:

- use of human cheek cells
- very hot agarose
- use of electrophoresis equipment
- use of ethidium bromide

Materials

Class equipment and supplies

microcentrifuge
clinical centrifuge
sterile saline solution, 0.9% NaCl
56°C water bath
95°C heat block (boiling water bath/hot plate)
10× PCR buffer
Taq DNA polymerase
10 mM deoxynucleoside triphosphates
TPA-25 primers (or APOC2 or D1S80)
thermal cycler
UV transilluminator
ethidium bromide staining solution

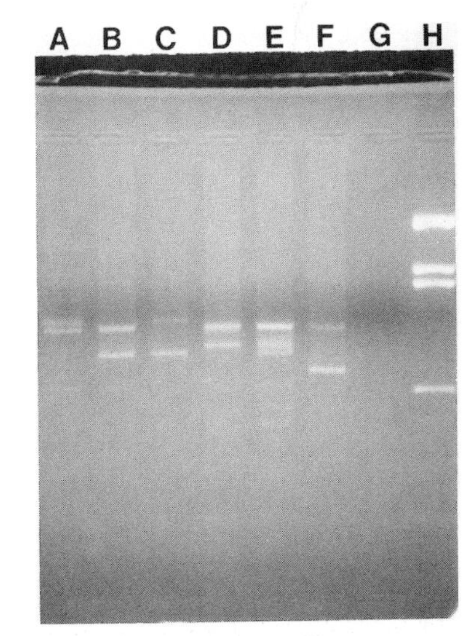

Figure 10.4 PCR of the D1S80 locus in a family. Lane A, student's sibling; lane B, student; lane C, father; lane D, mother; lane E, maternal grandmother; lane F, maternal grandfather; lane G, PCR master mix without template; lane H, pBR322 *Bst*NI digest DNA marker.

Polaroid camera and film
electrophoresis chambers and power supplies
1× Tris-borate-EDTA buffer for electrophoresis
mineral oil
microcentrifuge tubes
2% agarose gel

Materials for each student
plastic medicine cup
paper cup
500-μl aliquot of InstaGene beads
1 aliquot of PCR H buffer
1 aliquot of 10-mg/ml proteinase K
3 sterile microcentrifuge tubes
15-ml blue-capped conical tube
small PCR tube of PCR master mix
0.5- to 10.0-μl micropipette and sterile tips
10- to 100-μl micropipette and sterile tips
100- to 1,000-μl micropipette and sterile tips
marking pen
1 aliquot of loading dye
1 aliquot of PCR oil

Protocols

PROTOCOL 1 OBTAINING CHEEK CELLS

A brief rinse will collect the normally occurring "loose" epithelial cells from the surface of the buccal mucosa.

1. Fill the plastic medicine cup with about 10 ml of sterile salt solution (0.9% NaCl). Rinse your mouth with the salt solution for about 10 seconds. Expel the salt solution into a paper cup.

Keep your cup and label your tube!

Centrifugation will force the cells to the bottom of the tube.

2. Carefully pour the salt solution, now containing your cheek cells, into a 15-ml tube. Close the tube and centrifuge for 10 minutes.

3. Pour the supernatant back into the paper cup. Be careful not to disturb the cell pellet at the bottom of the tube.

4. Discard the paper cup containing the supernatant into the appropriate waste container.

PROTOCOL 2 DNA ISOLATION USING INSTAGENE BEADS

Breaking cell clumps completely exposes all cells to InstaGene reagents.

1. Resuspend the InstaGene solution with your pipette and add all of the InstaGene beads (500 μl) to your cells in the blue-capped tube. Pipette up and down to break up the clump of cells and mix them with the InstaGene beads.

2. Transfer the InstaGene beads and your cells back to the small tube. Use a vortex shaker to mix the cells and the beads.

Incubation at 56°C ensures cell lysis and gently denatures DNA from associated histones.

3. Incubate your tube at 56°C for 15 minutes.

Incubation at 95°C completely denatures proteins from the cell.

4. Vortex again; then heat the tube for 8 minutes at 95°C in the heating block.

5. Allow the tube to cool and then centrifuge for 3 minutes.

InstaGene beads chelate divalent ions, such as Mg^{2+}.

InstaGene bead reagents and cell debris will interfere with PCR.

6. Carefully remove most of the supernatant, containing your DNA, from the tube. Do not pipette any InstaGene beads or cell debris from the bottom of the tube. Transfer the liquid to a clean tube.

7. Discard the tube containing the pellet into the appropriate waste container.

8. DNA may be stored at –20°C for future exercises. Use 25 μl of this DNA for PCR analysis.

PROTOCOL 3 ISOLATION OF DNA FROM HAIR

1. Pipette 100 µl of PCR H buffer into a microcentrifuge tube.

PCR H buffer is very similar to the buffer used for PCR with the addition of the detergents NP-40 and Tween 20.

2. Add 2.5 µl of 10-mg/ml proteinase K solution to the tube.

NP-40 and Tween 20 assist with lysis of the cell and denaturation of cellular proteins.

3. Pluck two or three hairs from your head, eyebrow, or arm. Be sure each hair is complete, with the bulbous portion of the root attached.

Proteinase K also degrades cellular proteins.

4. Cut off about 5 mm of the hair at the root end and add it to the tube of PCR H buffer plus proteinase K.
5. Mix the hair and buffer by inverting the tube several times.
6. Pulse the tube in the microcentrifuge. Make sure your hair is in the buffer.
7. Incubate the tube at 56°C for 1 hour.

Proteinase K functions best at 56°C.

8. Heat your tube to 95°C in the heating block for 10 minutes.

Incubation at 95°C inactivates the proteinase K, which has the potential for denaturing the Taq *DNA polymerase in the PCR mixture.*

9. Allow your tube to cool to room temperature.
10. Use 25 µl of your DNA extract for PCR analysis.

PROTOCOL 4 THE PCR

PCR master mix will be made by the instructor.

1. Add 25 µl of DNA isolated with InstaGene beads to 25 µl of PCR master mix.

PCR master mix contains all four deoxynucleoside triphosphates, Taq *DNA polymerase, buffer with Mg^{2+}, and two different oligonucleotide primers.*

1a. *For APOC2,* use 15 µl of DNA and 25 µl of PCR master mix, add 10 µl of dH_2O to adjust the volume to 50 µl total, *and/or* add 25 µl of DNA isolated from hair to 25 µl of PCR master mix.

2. Write your number on top of each PCR tube.
3. Pulse the tubes in the microcentrifuge to mix.
4. Add 50 µl of PCR oil to each tube.

PCR oil is very pure mineral oil: a thin layer prevents evaporation during temperature cycling.

5. Place the samples in the thermal cycler.

Some thermal cyclers are equipped with a heated lid, which eliminates the need for oil.

Included in the thermal cycler will be a tube of PCR master mix without DNA template as a negative control.

Programming of the thermal cycler will be demonstrated.

Reaction Conditions for Amplification of TPA-25, APOC2, and D1S80

cycle 1	step 1	94°C for 5 minutes
	step 2	65°C for 2 minutes
	step 3	72°C for 2 minutes
cycles 2 to 30	step 1	94°C for 2 minutes
	step 2	65°C for 2 minutes
	step 3	72°C for 2 minutes
cycle 31	step 1	94°C for 2 minutes
	step 2	65°C for 2 minutes
	step 3	72°C for 5 minutes

Hold at room temperature or 4°C.

Cycle 1 contains an extended 94°C step to completely denature template DNA.

Cycle 31 contains an extended elongation step to permit completion of amplification products.

PROTOCOL 5 ANALYSIS OF PCR PRODUCTS

Analysis of PCR products is performed by gel electrophoresis. Agarose gels (2%) have been prepared by your instructor. Everyone's samples will be electrophoresed together on several gels for comparison.

2% agarose gels provide optimal resolution of PCR products less than 1,000 bp in size.

Oil from the PCR tube contaminating the sample will interfere with gel loading.

Loading dye makes the sample dense and provides a means of assessing the progress of electrophoresis.

One well in each gel will contain DNA size markers, and another will include the negative control.

1. Remove 20 µl of your amplified DNA sample from underneath the oil and pipette it into a clean microcentrifuge tube.
2. Add 2 µl of loading dye and pulse your tube in the microcentrifuge.
3. Carefully pipette your sample into one of the wells, being sure to note which well contains your sample.
4. Everyone's samples will be electrophoresed together, and a photograph will be taken of the gel results.

References and General Reading

Batzer, M. A., and P. L. Deininger. 1991. A human specific subfamily of Alu sequences. *Genomics* **9:**481–487.

Batzer, M. A., V. A. Gudi, J. C. Mena, D. W. Foltz, R. J. Herrera, and P. L. Deininger. 1991. Amplification dynamics of human-specific (HS) Alu family members. *Nucleic Acids Res.* **19:**3619–3623.

Budowle, B., R. Chakraborty, A. M. Giusti, A. J. Eisenberg, and R. C. Allens. 1991. Analysis of the VNTR locus D1 S80 by the PCR followed by high resolution PAGE. *Am. J. Hum. Genet.* **48:**137–144.

Budowle, B., B. W. Koons, and J. D. Errera. 1996. Multiplex amplification and typing procedure for the loci D1 S80 and amelogenin. *J. Forensic Sci.* **41:**660–663.

Campbell, A. M., J. Williamson, D. Padula, and S. Sundby. 1997. Use PCR and a single hair to produce a "DNA fingerprint." *Am. Biol. Teacher* **59:**172–178.

Erlich, H. A. 1992. *PCR Technology: Principles and Applications for DNA Amplification.* W. H. Freeman & Co. New York, N.Y.

Hochmeister, M. N., B. Budowle, J. Jung, U. V. Borer, C. T. Comey, and R. Dirnhofer. 1991. PCR based typing of DNA extracted from cigarette butts. *Int. J. Leg. Med.* **104:**229–233.

Kirby, L. T. 1992. *DNA Fingerprinting: an Introduction.* W. H. Freeman & Co., New York, N.Y.

Kloosterman, A. D., B. Budowle, and P. Daselaar. 1993. PCR amplification and detection of the human DlS80 VNTR locus. Amplification conditions, population genetics and application in forensics analysis. *Int. J. Leg. Med.* **105:**257–264.

Mullis, K. B. 1990. The unusual origin of the polymerase chain reaction. *Sci. Am.* **262:**56–65.

Mullis, K. B., F. A. Faloona, S. J. Scharf, R. K. Saiki, G. T. Horn, and H. A. Erlich. 1986. Specific enzymatic amplification of DNA in vitro: the polymerase chain reaction. *Cold Spring Harbor Symp. Quant. Biol.* **51:**263–273.

Sjantilla, A., B. Budowle, M. Strom, V. Johnson, M. Lukka, L. Peltonen, and C. Ehnholm. 1992. PCR amplification of alleles at the D1 S80 locus: comparison of a Finnish and a North American Caucasian population sample and forensic casework evaluation. *Am. J. Hum. Genet.* **50:**816–825.

Discussion

1. Why is the denaturation step performed in the first cycle of PCR longer than the denaturation steps in the rest of the cycles?
2. What effect might increasing the annealing temperature have on PCR?
3. Why is there a 31st cycle with a longer elongation time than the other cycles?

Examine the photograph of your PCR results. Locate the DNA marker lane, the negative control lane, and the lane containing your results.

4. What do you observe in the DNA marker lane? What is its purpose?
5. What is the negative control lane? What do you observe? What is its purpose?

6. If you see bands of DNA in the negative control lane, what do you think this means?
7. Compare the lane containing your results with the negative control lane. Do you observe any similarities? Do you observe any differences? What do you think this means?
8. Which bands of DNA represent your amplified locus?
9. Are you homozygous or heterozygous at this locus?
10. Compare the lane containing your results with the lanes of other individuals. Do you find any similarities? Do you observe any differences? What do you think this means?
11. If you can't identify any amplified bands, what are potential sources of error in your experiment?
12. Do you observe more than two DNA bands in any lane? Why do you think this is so?
13. Examine Fig. 10.4 from PCR of the D1S80 locus in samples obtained from a family. Can you trace the inheritance of the D1S80 bands from parent to child?

EXERCISE 11

Quantitation of DNA by Spectrophotometry

Key Concepts

1. Pure DNA has an absorption maximum at 260 nm.
2. The quantity of DNA present in a sample can be determined by measuring absorbance at 260 nm.
3. The ratio of absorbance at 260 nm to absorbance at 280 nm (A_{260}/A_{280}) can be used to indicate DNA purity.
4. DNA sample contamination is indicated by absorbance at 230 and 325 nm.

Goals

By the end of this exercise, you will be able to

- prepare a standard curve of control DNA concentrations
- measure absorbance at 230, 260, 280, and 325 nm
- calculate the quantity of DNA in unknown samples
- determine the purity of DNA in unknown samples

Introduction

Researchers often must determine the quantity and quality of DNA present in a sample. When extracting DNA, one must be able to ascertain the yield of DNA per cell or per sample or to compare yields obtained by different extraction protocols. Researchers must judge the quantity and quality of DNA for use in cloning, transfection, Southern blotting, restriction enzyme analysis, sequencing, polymerase chain reaction, and transformation. Too much or too little DNA can interfere with results. DNA samples containing contaminants may not be pure enough for some applications.

There are several methods to determine the quantity and concentration of DNA in a sample. These include absorption spectrophotometry (Fig. 11.1), fluorescence spectrophotometry, and agarose gel electrophoresis. Although fluorescence spectrophotometry is the most sensitive, detecting DNA at concentrations of 0.01 to 15 μg/ml, absorption spectrophotometry is still the standard method. Absorption spectrophotometry detects 1 to 50 μg of DNA per ml. Quantitation of DNA by agarose gel electrophoresis provides a greater degree of sensitivity for very small samples and is discussed in exercise 12.

Each of these methods has advantages and disadvantages. Absorption spectrophotometry is a reliable method for assessing both the quantity and purity of DNA, because DNA, RNA, protein, and other possible contaminants absorb in the ultraviolet (UV) wavelength range. However, this method requires larger quantities of DNA to obtain accurate readings (e.g., 500 ng of DNA per ml yields an A_{260} of 0.01) and cannot distinguish between DNA and RNA. Fluorescence spectrophotometry (using the dye Hoechst 33258) does not measure RNA

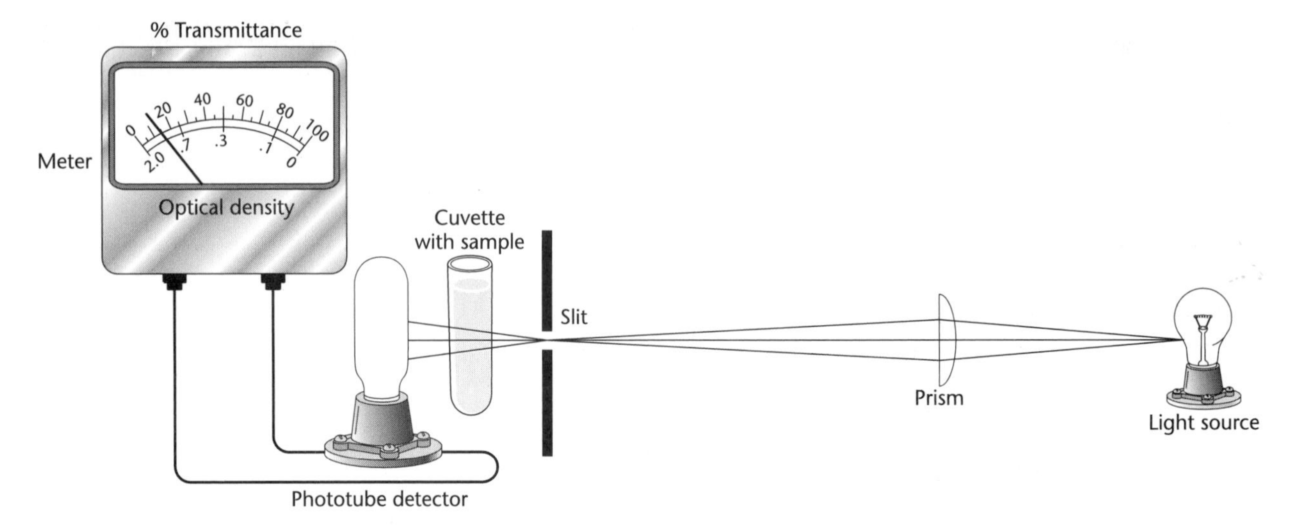

Figure 11.1 A simple diagram of a spectrophotometer. Light, either UV or visible spectrum, passes through a prism and an adjustable slit. This permits selection of a specific wavelength to pass through the cuvette containing the sample to be analyzed. The light passing through the sample is collected at the phototube detector, converted into electrical current, and transmitted to the meter, where results can be measured either as percent transmittance or as optical density (absorbance).

and therefore is specific for DNA, but because the dye binds preferentially to adenine-thymine (AT) sequences, its use is more limited when the DNA studied possesses a high guanosine-cytosine (GC) content. An advantage of fluorescence spectrophotometry is that DNA concentration in cell lysates and other crude preparations can be determined.

The absorption properties of DNA can be used to detect, quantify, and estimate purity of DNA samples. DNA absorbs UV light maximally at 260 nm owing to the chemical nature of the nitrogenous base moieties. Deoxyribose and phosphate groups do not contribute appreciably to absorption. A_{260} values quantitate pure DNA in the microgram range. Because RNA absorbs maximally at A_{260} as well, this method cannot discriminate between DNA and RNA. But the method can also be used to quantitate pure RNA. An A_{260} of 1.0 indicates 50 µg of double-stranded DNA per ml, or approximately 37 µg of single-stranded DNA per ml, or approximately 40 µg of single-stranded RNA per ml.

Absorbance is read at 325, 280, 260, and 230 nm to quantitate DNA and determine a variety of contaminants. The purity of DNA is critical to many biotechnology experiments. Contaminating cellular proteins may bind to target DNA, interfering with enzymes required for an experiment, such as restriction enzymes or DNA polymerases. This interference can invalidate an experiment. Absorbance at 325 nm is often due to particulates in the sample solution or dirty cuvettes. Aromatic moieties such as phenol, and small molecules containing peptide bonds, absorb at 230 nm. Proteins have a peak absorption at 280 nm, and their presence in a sample will reduce the A_{260}/A_{280} ratio. An A_{260}/A_{280} ratio of 1.8 to 1.9 indicates very pure DNA, and an A_{260}/A_{280} ratio or 1.9 to 2.0 indicates very pure RNA. Lower ratios indicate protein contamination.

In this exercise you will prepare a standard curve using known concentrations of DNA, and determine the concentration of DNA in unknown samples by comparing absorbance of the unknown samples against the standard curve.

Lab Language

A_{260}/A_{280}
absorbance
absorption
cuvette
fluorescence
spectrophotometer
spectrophotometry
wavelength

Laboratory Safety

Review the sections on physical, biological, and chemical hazards in "Laboratory Safety." Special considerations for this exercise:

UV spectrophotometer

Materials

Class equipment and supplies
UV spectrophotometer
quartz cuvettes
10× TNE (Tris-sodium chloride-EDTA) buffer

Students will need to prepare
1× TNE buffer

Materials per pair of students
tubes for preparing dilutions
2 quartz cuvettes, matched
calf thymus DNA standard solutions
DNA (unknown)
1-ml serological pipette and pipette aid
10-ml serological pipette and pipette aid
disposable 5-ml test tubes
50-ml graduated cylinder
50-ml or 100-ml medium bottle

Protocols

PROTOCOL 1 PREPARATION OF STANDARD CURVE DILUTIONS

1. Prepare 50 ml of 1× TNE buffer. Determine the quantity of 10× TNE necessary. Document your calculations below.

2. Add approximately 20 ml of dH_2O to a 50-ml graduated cylinder. Using a 10-ml pipette and pipette aid, add the required volume of 10× TNE to the graduated cylinder. Bring to volume with dH_2O.

3. Prepare dilutions of known, standard DNA in 1× TNE for the standard curve. Choose a final volume appropriate for use in your spectrophotometer. Use the chart below to document your calculations.

Concentrations between 1.0 and 60 µg of DNA per ml should provide an acceptable range.

Concentration of DNA (µg/ml)	Volume of DNA standard	Volume of TNE	Final volume
1.0			
5.0			
10.0			
20.0			
30.0			
40.0			
50.0			
60.0			

PROTOCOL 2 PREPARATION OF UNKNOWN SAMPLES

1. Obtain a sample of DNA of unknown concentration in 1× TNE buffer.

2. Prepare the following dilutions in 1× TNE: 1:2, 1:5, 1:10, and 1:20. Document your calculations in the chart below.

It may be necessary to dilute the unknown sample to provide a concentration that will yield an absorbance reading falling within the standard curve.

Dilution	Volume of unknown DNA	Volume of TNE buffer
1:2		
1:5		
1:10		
1:20		

PROTOCOL 3 SPECTROPHOTOMETRIC MEASUREMENT AND CONSTRUCTION OF STANDARD CURVE

Many spectrophotometers must be warmed up for 15 to 20 minutes before use.

Wavelengths in the UV range will pass through quartz, but not through glass.

1. Turn on the spectrophotometer. Set the wavelength to 325 nm (A_{325}).
2. Obtain two matched quartz cuvettes.
3. Without a cuvette in the sample holder, close the cover and set optical density to infinity.
4. Pipette the manufacturer's recommended volume of TNE into one cuvette. This is the TNE blank.

To zero the instrument means to set the absorbance at 0.

5. Insert the cuvette containing TNE into the UV spectrophotometer set at 325 nm, and zero the instrument.

Handle cuvettes only at the top edge.

Fingerprints on the cuvettes interfere with absorbance readings.

6. Place the first DNA standard concentration into the second cuvette, insert it into the spectrophotometer, and read absorbance. Record your results in Table 11.1.
7. Return the TNE blank to the spectrophotometer, and readjust to zero if necessary.
8. Repeat steps 6 and 7 for each concentration of standard DNA.
9. Repeat the protocol, steps 3 through 8, at A_{280}, A_{260}, and A_{230}.

Table 11.1 Absorbance data for standard curve[a]

Concentration of DNA (μg/ml)	Absorbance 325 nm	Absorbance 280 nm	Absorbance 260 nm	Absorbance 230 nm
1.0				
5.0				
10.0				
20.0				
30.0				
40.0				
50.0				
60.0				

[a]An absorbance of 1.0 at 260 nm is equivalent to 50 μg of double-stranded DNA per ml, approximately 37 μg of single-stranded DNA per ml, or approximately 40 μg of single-stranded RNA per ml.

Ratios of 1.8 to 1.9 indicate very pure DNA.

Protein absorbs at 280 nm and will lower the ratio.

Dirty cuvettes and particulate matter will give absorbance readings at 325 nm.

Interference from urea or phenol contaminants will give absorbance at 230 nm.

10. Construct a graph using the data for A_{260}. Plot absorbance on the *y* axis and DNA concentration on the *x* axis. You will use this curve to extrapolate the concentration of DNA in unknown samples.
11. Estimate the purity of the standard DNA by determining the A_{260}/A_{280} ratio.

PROTOCOL 4 QUANTITATION OF DNA IN UNKNOWN SAMPLES

1. Insert the cuvette containing TNE into the UV spectrophotometer set at 325 nm, and zero the instrument.

2. Place the first unknown DNA dilution into the second cuvette, insert it into the spectrophotometer, and read absorbance. Record your results in Table 11.2.

3. Return the TNE blank to the spectrophotometer, and readjust to zero if necessary.

4. Repeat steps 2 and 3 for each dilution of unknown DNA.

5. Repeat the protocol, steps 1 through 4, at A_{280}, A_{260}, and A_{230}.

Table 11.2 Absorbance data for unknown samples

Dilution of DNA	Absorbance			
	325 nm	280 nm	260 nm	230 nm
1:2				
1:5				
1:10				
1:20				
Undiluted				

6. Using absorbance, calculate the concentration of DNA in your unknown sample dilutions by extrapolating from your standard curve. Account for your dilution factor.

Absorbance readings below 0.1 and above 1.0 usually are not considered accurate.

7. Determine the A_{260}/A_{280} ratio of each dilution to estimate the purity of your sample.

Reference

Ausubel, F. M., R. Brent, R. E. Kingston, D. D. Moore, J. G. Seidman, J. A. Smith, and K. Struhl (ed.). 1994. *Current Protocols in Molecular Biology.* Wiley Interscience, New York, N.Y.

Discussion

1. Why is it possible to quantify DNA using absorbance?
2. Discuss the relative absorbance of DNA, RNA, and protein.
3. How can you determine whether a DNA sample is pure?
4. Does the calculated purity of your DNA standards agree with the assay data provided by the supplier? Interpret any differences you might encounter.
5. Discuss the quantity of DNA in your unknown sample and its purity.
6. Do your determinations of DNA concentration in the various dilutions of your unknown samples agree? Why or why not?

EXERCISE 12

Quantitation of DNA by Gel Electrophoresis

Key Concepts

1. Quantities of DNA fragments used in cloning experiments are too small to be determined by spectrophotometry.
2. One can determine DNA quantities by comparing an unknown DNA with a known quantity of DNA of similar size.

Goals

By the end of this exercise, you will be able to

- precisely load agarose gels and perform gel electrophoresis
- determine the quantity of DNA in an unknown sample

Introduction

Agarose gel electrophoresis may be used to estimate the quantity of DNA in unknown samples. This is especially useful when the sample is very small (100 μl) or when the concentration is low (5 μg/ml). By electrophoresing marker DNA fragments of known sizes, one may compare an unknown DNA sample on an agarose gel with the markers and estimate DNA quantity and concentration (Fig. 12.1).

Small volumes and quantities of DNA samples are frequently used in the biotechnology laboratory. For example, as little as 50 ng of plasmid DNA in a volume of 10 μl is sufficient to transform bacteria. Only 2 or 3 μg of plasmid DNA, needed for analysis of a cloning experiment, may be recovered from a culture of transformed bacteria. For these and many other applications, researchers must be able to estimate very small DNA quantities.

A standard method for quantifying DNA is by spectrophotometry, as discussed in exercise 11. For spectrophotometric determinations, the diluted sample volume is usually 0.5 or 1.0 ml and requires a concentration of 5 μg of DNA per ml to be accurate. DNA may not be available in this volume and concentration. DNA samples may be isolated in volumes as small as 100 μl, and diluting them may not be appropriate for a proposed experiment.

When quantitating DNA by agarose gel electrophoresis, one must work very carefully and precisely. Dilutions and gel loading must be performed accurately to ensure good results. Usually, 1 μg of a DNA marker containing known fragment sizes, to provide controls, is electrophoresed into an agarose gel concurrent with the unknown DNA sample. The DNA marker control must include DNA fragments of approximately the same size as the unknown DNA. The agarose gel is stained with ethidium bromide, and the intensity of the unknown DNA sample is compared with a known band of equivalent size. Stain intensity is proportional to the amount of DNA present. One must carefully consider the fragment size, since intensity by the human eye can be deceptive.

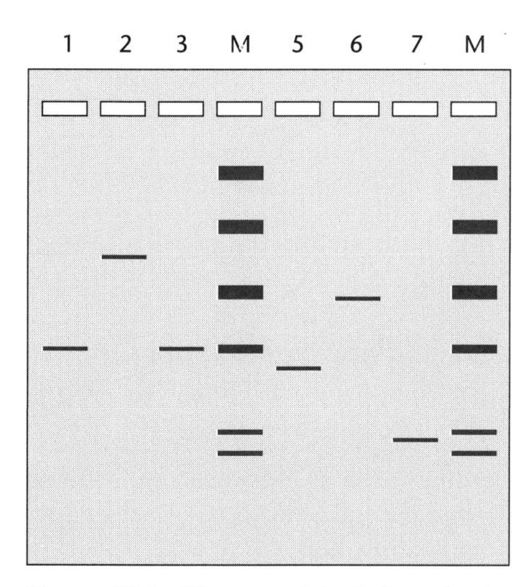

Figure 12.1 Diagram of typical results for quantitating DNA by gel electrophoresis. Lanes 1 to 3 and 5 to 7 are unknown samples. Lane M represents a *Hin*dIII digest of bacteriophage lambda (λ) DNA, a standard DNA marker.

EXERCISE 12

Lab Language

agarose
aliquot
anode/positive pole
cathode/negative pole
electrophoresis

Laboratory Safety

Review the sections on physical, biological, and chemical hazards in "Laboratory Safety." Special considerations for this exercise:

boiling hot agarose
electrophoresis equipment
ethidium bromide
hot plate

Materials

Class equipment and supplies
microcentrifuge
0.8% agarose
10× Tris-borate-EDTA (TBE) buffer (or from exercise 3)
masking tape
dH_2O
DNA marker (lambda DNA; *Hin*dIII digest of lambda DNA)
ethidium bromide solution
ultraviolet (UV) transilluminator
Polaroid camera and black-and-white type 667 film

Students will need to prepare
300 ml of 1× TBE
Tris-EDTA (TE) buffer

Materials per pair of students
gel electrophoresis chamber, gel casting tray, comb
power supply (two pairs of students will share this)
0.5- to 10.0-µl micropipette and sterile tips
1 aliquot of DNA marker
1 aliquot of unknown DNA
1 aliquot of loading dye
sterile microcentrifuge tubes
microcentrifuge tube rack
disposable exam gloves
permanent marking pen
weigh boat

Protocols

PROTOCOL 1 PREPARATION OF SAMPLES

ADD 10 µl of lambda marker to gel

1. Obtain a DNA marker control of known quantity, fragment sizes, and concentration. Dilute it to 9 µl total with 1× TE. Add 1 µl of loading dye.

2. Obtain a test DNA sample of known volume. Dilute the sample to 9 µl total with 1× TE and add 1 µl of loading dye.

3. Pop spin the samples to mix and pool the reagents in the bottom of the tube.

PROTOCOL 2 CASTING THE AGAROSE GEL AND PREPARING 1× TBE BUFFER

A. Cast the Gel

1. Seal the ends of the casting tray according to the manufacturer's directions and insert the comb in the notch closest to the end of the tray. Set the casting tray on your lab bench where it will not be disturbed.

0.8% agarose provides the optimal pore size for resolution of the unknown DNA.

2. Obtain warm melted 0.8% agarose solution and pour it into the casting tray to a depth of about 5 mm. Agarose will appear cloudy when solidified, about

10 to 30 minutes. Do not move the casting tray during solidification. Imperfections in the gel will skew your results.

B. Prepare 1× TBE Running Buffer

1. Prepare 300 ml of 1× TBE buffer. Calculate how much 10× TBE stock to use.

TBE is used to conduct the current through the gel.

2. Using a 50-ml graduated cylinder, measure the appropriate amount of 10× TBE and pour it into a 500-ml graduated cylinder.
3. Dilute the 10× TBE with distilled or deionized water to a final volume of 300 ml.

Do not use tap water: minerals will interfere with electrophoresis.

C. Prepare the Gel and Gel Chamber

1. When the agarose has solidified, unseal the ends of the casting tray and place it in the gel chamber. Make sure the comb is at the cathode, the negative pole.

DNA is negatively charged and will migrate toward the anode.

2. Fill the box with just enough 1× TBE buffer to cover the surface of the gel.
3. *Gently* remove the comb, taking care not to tear the gel. Make sure the wells formed by the comb are completely submerged. Add more buffer if necessary. The gel is now ready to load with the DNA samples.

PROTOCOL 3 LOADING THE GEL AND SETTING UP THE GEL CHAMBER FOR ELECTROPHORESIS

1. Carefully pipette 10 µl from each sample into separate wells in the agarose gel. Use a fresh tip for each sample. Correlate and record sample identification with well number.
2. Slide the cover of the electrophoresis chamber into place and connect the electrical leads to the power supply. Make sure your leads are connected so that the DNA migrates the correct direction in the gel.

Red to red; black to black.

On most gel boxes, the anode is red and the cathode is black.

3. Turn on the power supply. Electrophorese your gel at 150 volts. You will run your gel until the purple dye front is near the end of the gel; observe at 10-minute intervals.

A stream of fine O_2 bubbles appears at the positive electrode (anode); H_2 gas bubbles at the negative electrode (cathode), indicating current flow.

4. When the electrophoresis is complete, after approximately 40 to 60 minutes, **turn off the power supply.** Disconnect the leads from the power supply and remove the top of the gel chamber.

Shortly after the current is applied, loading dye can be seen migrating through the gel toward the positive electrode.

5. Discard the buffer in the sink.

Bromophenol blue (blue dye) migrates at about the same rate as 2,800-bp fragments of DNA.

Xylene cyanol (purple dye) migrates similarly to 250-bp fragments of DNA.

PROTOCOL 4 ETHIDIUM BROMIDE STAINING OF DNA IN AGAROSE GELS

1. **Wearing disposable vinyl or latex exam gloves,** remove the casting tray from the gel chamber. *The gel is very slippery and fragile. Hold the casting tray level.* Slide the gel, right side up, into a disposable plastic weigh boat or other suitable tray. Add sufficient 1-µg/ml ethidium bromide solution to cover the gel.

DNA is colorless; thus it is not immediately visible in a gel.

2. Stain the gel for 5 to 10 minutes at room temperature.

Ethidium bromide is a mutagen and suspected carcinogen.

3. Decant the ethidium bromide back into the stain bottle; it may be used to stain several gels.

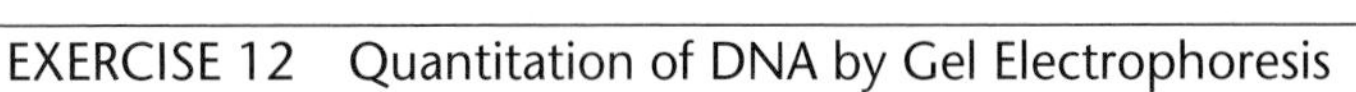

Fluorescence intensity is directly proportional to the amount of DNA present.

UV light can damage skin and eyes.

Some UV transilluminators have a UV-blocking safety shield.

4. Destain in dH_2O for 5 to 10 minutes.

5. Visualize the DNA bands by placing the gel on a UV light transilluminator. Ethidium bromide binds to DNA and fluoresces orange. **If your UV transilluminator does not have a UV-blocking lid, wear UV-blocking eye protection: either a face mask or goggles.**

6. Turn on the transilluminator and photograph the gel using a Polaroid camera and black-and-white type 667 film. Use an orange filter. Approximate settings are f-stop 8, 1/2-second exposure.

7. Discard the gel and all other materials contaminated with ethidium bromide in the appropriate waste container.

PROTOCOL 5 ESTIMATION OF UNKNOWN DNA QUANTITY

1. Use the following chart for your calculations. Write in the name of the DNA marker used, and determine the total quantity, in micrograms, loaded into the gel.

DNA marker =		Quantity of marker DNA (μg) =	
Band number	**Size (bp)**	**% of total bp**	**Quantity of DNA (μg)**
1			
2			
3			
4			
5			
6			
7			
8			
9			
Total size (bp)		100%	

2. Fill in the DNA size fragments contained in this marker.

3. Add the DNA fragment sizes to determine the total base-pair size of the marker.

4. Calculate the percentage of DNA base pairs in each band by dividing the band fragment size by the total DNA size.

5. Determine the total amount of DNA in each band by multiplying the percentage of DNA base pairs by the total quantity of DNA loaded into the gel.

6. Estimate the quantity of DNA in the unknown sample by comparing the intensity of the unknown band with the intensity of the DNA marker band closest in size.

Reference

Sambrook, J., E. F. Fritsch, and T. Maniatis. 1989. *Molecular Cloning: a Laboratory Manual.* Cold Spring Harbor Laboratory Press, Cold Spring Harbor, N.Y.

Discussion

1. Compare your determination with the actual concentration of the unknown DNA quantity. Discuss the accuracy of your determination.

2. How could more accurate results be obtained?
3. List and discuss five similarities of DNA quantitation by spectrophotometry and gel electrophoresis.
4. List and discuss five differences between DNA quantitation by spectrophotometry and gel electrophoresis.

Applying the Fundamentals

SECTION III

DNA Cloning

EXERCISE 13

DNA Ligation and Cloning

Key Concepts

1. Cloning a desired piece of DNA into a plasmid vector, then transforming a suitable bacterial host with the recombinant plasmid, is a very effective way of replicating large amounts of DNA for a variety of uses.
2. Plasmid vectors used for cloning have been engineered by scientists from plasmids that naturally occur in bacteria. These new vectors contain an origin of replication, a selectable marker (usually antibiotic resistance), and a multiple cloning site.
3. To screen for successful insertion of desired DNA into a plasmid, the plasmid can contain an additional selectable marker. Disruption of the latter gene changes phenotypic characteristics of the bacteria.
4. DNA ligase is the enzyme that catalyzes the formation of phosphodiester bonds to ligate DNA fragments together. DNA ligase may join sticky-ended or blunt-ended fragments of the desired DNA and plasmid DNA together.

Goals

By the end of this exercise, you will be able to

- identify useful features in a cloning vector
- discuss the process of DNA ligation using sticky ends
- describe strategies for screening for a desired DNA insert

Introduction

Cloning a new piece of DNA has almost countless uses. These uses range from determining the exact nucleotide sequence of a gene or piece of DNA to obtaining the protein product of a gene for use as a medication to gene therapy.

To clone a piece of DNA, one must have a desired piece of DNA to clone. This may involve finding a gene that a researcher *only suspects* exists. Sometimes cloning a piece of DNA may simply involve moving a gene from one plasmid or vector to a different vector that has special properties. Where and how do you start to identify new genes? How do you choose a new vector? What special characteristics are desirable in a cloning vector?

DNA Libraries

One way to obtain desired DNA is to make and screen a library (exercise 29). A DNA library is a collection of yeast or bacteria clones. These clones contain recombinant vectors carrying DNA fragments from the genome of a single species. This is one potential source of new pieces of DNA. The library can be a genomic library, containing all of the sequences of an organism, including introns as well as exons, or it can be a complementary DNA (cDNA) library, containing only the expressed exons. The challenge comes in finding a desired gene sequence in a library. Sometimes a library is screened with a gene that has already been identified in another organism. This enables the researcher to identify and compare genes in various species. Another method of screening a library is to express the protein products of the genes and screen with an antibody specific for the desired protein.

Occasionally, a piece of DNA is moved from one vector to another vector. In this case, the gene of interest has already been isolated and identified. The advantage of moving DNA from one vector to another depends on the specific characteristics of the vector and what a researcher desires to study about a given gene sequence. For example, in this experiment we will move the human growth hormone (hGH) from pBR322 into a suitable cloning vector, pGEM-3Zf+, which can be used to produce single-stranded DNA or for in vitro transcription studies.

pGEM-3Zf(+/−)

One special set of vectors, which contains an additional selection characteristic and can be used for in vitro transcription or to produce single-stranded DNA is pGEM-3Zf(+/−). There are two vectors in this set, pGEM-3Zf+ and pGEM-3Zf−. The two vectors are identical except for the orientation of the filamentous bacteriophage f1 origin of replication. Using the f1 origin in each orientation provides a mechanism for producing each DNA strand in its pure single-stranded form for sequencing by the dideoxy sequencing method or for mutagenesis studies. Bacteria containing a pGEM-3Zf(+/−) vector are infected with a helper phage that causes the plasmid to replicate single-stranded DNA and to package it in viruslike particles.

In this exercise, we will be using pGEM-3Zf+, abbreviated here as pGEM+. In addition to the origin of replication, the gene for ampicillin resistance, and the multiple cloning site (MCS), pGEM+ contains the MCS within the *lac* operon, adjacent to the *lacZ* gene (Fig. 13.1 and 13.2). The *lacZ* gene codes for β-galactosidase (β-Gal). X-Gal (5-bromo-4-chloro-3-indolyl-β-D-galactopyranoside) is the substrate for β-Gal; cleavage of X-Gal results in a blue color. When *Escherichia coli*

Figure 13.1 The mechanism of blue or white color selection to screen for plasmids that contain insert DNA. (Left) IPTG induces the *lac* operon, turning on the *lacZ* gene that produces β-Gal. X-Gal, present in the bacterial growth medium, is cleaved by β-Gal, and bacterial colonies turn blue. (Right) A DNA fragment, inserted into the multiple cloning site (MCS) of the plasmid, disrupts the *lacZ* gene, and no β-Gal is produced. In this case, X-Gal remains intact and bacterial colonies are white, indicating the presence of plasmids containing inserts in these bacteria.

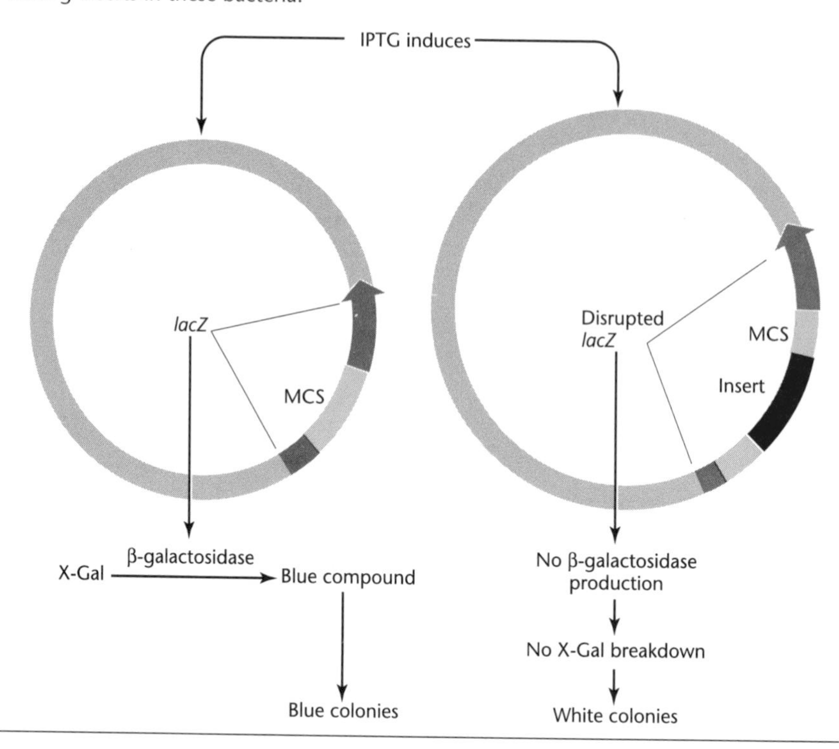

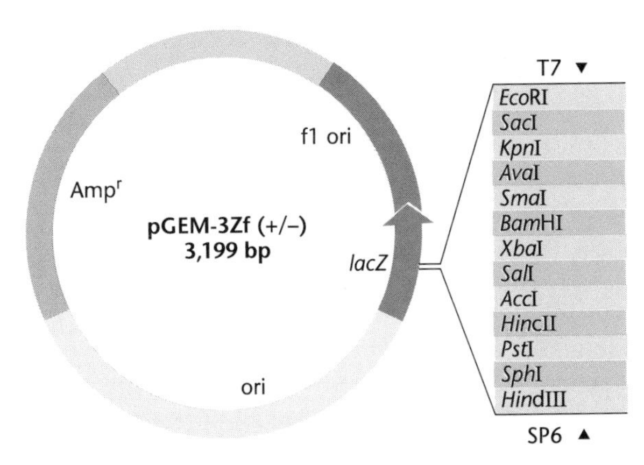

Figure 13.2 The cloning vector pGEM-3Zf+.

JM109 containing pGEM+ is grown on indicator plates containing the *lac* operon inducer IPTG (isopropyl-β-D-thiogalactopyranoside) and X-Gal, the bacterial colonies turn blue. IPTG "turns on" the *lac* operon, β-Gal is transcribed and translated from *lacZ*, and β-Gal metabolically cleaves X-Gal. The importance of having the *lacZ* gene next to the MCS is illustrated in Fig. 13.1. When a piece of DNA is inserted into the MCS, it results in disruption of the *lacZ* gene. Functional β-Gal is not produced, the X-Gal substrate is not cleaved, and the bacterial colonies are white, not blue. This provides a mechanism for selecting bacteria that are replicating plasmids that have inserts.

Ligation

T4 DNA ligase is an enzyme that will join two double-stranded fragments of DNA together. It requires a 3′ hydroxyl group on one fragment and a 5′ phosphate on the other to complete the synthesis of a covalent phosphodiester bond between the two fragments of DNA.

For ligation of a new fragment of DNA into a vector, the ends of the vector and the insert fragment of DNA must be compatible. This occurs when both are blunt-ended, having no overhangs, or when both possess complementary cohesive ends, also called sticky ends. Cohesive or blunt ends are generated by restriction enzyme digestion (see exercise 7). A specific restriction enzyme generates a characteristic and sequence-specific blunt or cohesive end. When blunt-ended DNA fragments and vectors are to be ligated together, the same restriction enzyme need not be used to generate the pieces. Blunt-ended fragments associate by random collision. After ligation, a restriction enzyme site is not regenerated. Sticky-ended DNA ligation involves complementary base pairing of vector and insert DNA overhangs. Using the same restriction enzyme to cleave both insert and vector DNA generates complementary sticky ends; after ligation, the same restriction enzyme site is regenerated. Restriction enzymes have been identified that recognize different sequences but generate the same overhang. Usually, a useful restriction enzyme site is not re-created. These are examples of nondirectional cloning because the insert DNA can be ligated into the vector in either orientation.

Directional cloning involves the use of two different restriction enzymes. One enzyme cuts at the 5′ end of the insert DNA, and the other cuts at the 3′ end of the insert. Similarly, the vector contains both restriction enzyme sites close together in the MCS. Double cutting of both insert and vector permits construction of recombinant DNA with the insert in a specific orientation.

When cohesive-ended ligations are performed, a 3:1 molar ratio of insert to vector is used. Therefore, 300 ng of insert plus 100 ng of vector would be ligated together if both are of equal length. However, if the plasmid is three times the length of the insert, then the amounts change to 100 ng of insert plus 100 ng of vector. With a directional cloning ligation, the ratio of insert to vector is less critical, and lesser amounts of insert can be used effectively.

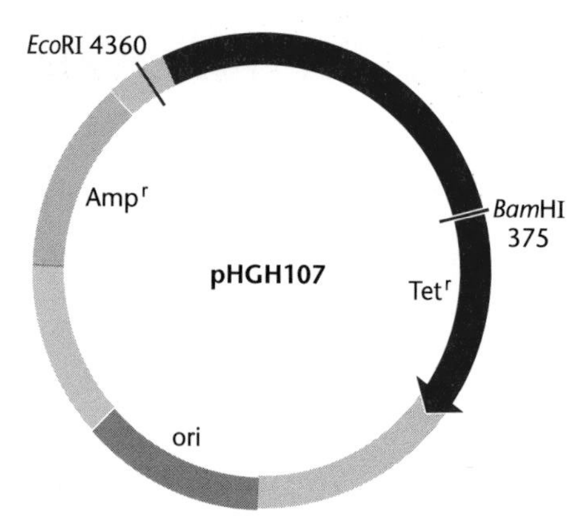

Figure 13.3 Plasmid pHGH107, containing the gene for hGH.

Human Growth Hormone

hGH is a peptide hormone consisting of slightly less than 200 amino acids. It is synthesized and released by the anterior pituitary gland. hGH has a number of effects, including both short-term metabolic alterations and long-term promotion of growth. Production peaks during adolescence when accelerated growth occurs. Like many other hormones, such as estrogen, progesterone, and melatonin, hGH production declines with age: people of age 70 to 80 are most likely to become hGH deficient. hGH is prescribed to prevent hypopituitary dwarfism in children suffering from growth hormone deficiency.

hGH was one of the first hormones produced by using recombinant DNA technology. Before recombinant hGH, growth hormone needed for therapeutic use was obtained from cadavers. The production of recombinant hGH revolutionized the pharmaceutical industry, producing a safer hormone free of possible contamination by infectious agents.

The hGH gene has been cloned into a carrying vector called pHGH107 (Fig. 13.3 and 13.4). A carrying vector is used to maintain the plasmid with its inserted DNA of interest in the bacterial host. pHGH107 contains genes for both ampicillin and tetracycline resistance as well as recognition sites for *Eco*RI and *Bam*HI. The hGH gene is positioned between these two recognition sites. Cleavage with both enzymes yields the gene insert with a different overhang at each end, thus permitting directional cloning.

Lab Language

β-Gal
blunt ends
cohesive ends
directional cloning
expression vector
IPTG
lac operon
lacZ
library
ligase
ligation
MCS
origin of replication
pGEM-3Zf(+/−)
plasmid
selectable marker
sticky ends
T4 DNA ligase
vector
X-Gal

Figure 13.4 The nucleotide sequence of the hGH gene.

Human Growth Hormone Sequence
Gene Bank Report

LOCUS A15074 651 bp
DEFINITION H.sapiens gene for growth hormone (hGH).
ACCESSION **A15074**

BASE COUNT 135 a 159 c 208 g 149 t

```
  1 gaagccacag ctgccctcca cagagcggca ctgcacgatg cgcaggaatg tctcgacctt
 61 gtccatgtcc ttcctgaagc agtagagcag cccgtagttc ttgagtagtg cgtcatcgtt
121 gtgtgagttt gtgtcgaact tgctgtaggt ctgcttgaag atctgcccag tccgggggct
181 gccatcttcc agcctcccca tcagcgtttg gatgccttcc tctaggtcct ttaggaggtc
241 atagacgttg ctgtcagagg cgccgtacac caggctgttg gcgaagacac tcctgaggga
301 ctgcacgggc tccagccacg actggatgag cagcagggag atgcggagca gctctaggtt
361 ggatttctgt tgtgtttcct ccctgttgga gggtgtcgga atagactctg agaaacagag
421 ggaggtctgg gggttctgca ggaatgaata cttctgttcc tttgggatat aggcttcttc
481 aaactcctgg taggtgtcaa aggccagctg gtgcagacga tgggcgcgga gcatagcgtt
541 gtcaaaaagc ctggataagg gaatggttgg gaaggcactg ccctcttgaa gccagggcag
601 gcagagcagg ccaaaagcca ggagcaggga cgtccgggag cctgtagcca t
```

Laboratory Safety

Review the sections on physical, biological, and chemical hazards in "Laboratory Safety." Special considerations for this exercise:

- hot agarose
- gel electrophoresis
- ethidium bromide staining
- use of microbes
- use of ultraviolet (UV) transilluminator
- use of X-Gal and IPTG

Laboratory Sequence

This exercise will require four to six laboratory sessions of 3 to 4 hours each, depending on the skill level of the students, allotted class time, and the extent of experimental analysis after cloning.

Day 1: Restriction enzyme digestion of pGEM-3Zf+ and pHGH107 (protocols 1 and 2)
Digestions may be frozen after incubation.
Preparation of agarose gel and buffer for electrophoresis (protocol 3)
Store gel wrapped in plastic wrap at 4°C or covered with buffer at room temperature.

Day 2: Gel electrophoresis (protocol 4)
This may be accomplished on day 1; store gels or slices at 4°C.
DNA purification (protocol 5)
Ligation reaction (protocol 6) (overnight in the refrigerator)

Day 3: Bacterial transformation and plating (protocols 7 and 8)

Day 4: Obtain and analyze results.
Pick colonies for further analysis (optional).

Days 5 and 6: Miniprep and analysis of isolated plasmids (optional)

Materials

Standard Materials for the Exercise

- 0.5- to 10-µl micropipette and sterile tips
- 10- to 100-µl micropipette and sterile tips
- 100- to 1,000-µl micropipette and sterile tips
- microcentrifuge
- sterile microcentrifuge tubes
- microcentrifuge test tube rack
- wax or permanent marking pen
- gloves

DNA Digestion and Agarose Gel Electrophoresis

Class equipment and supplies

- 37°C water bath or heat block
- 80°C water bath or heat block
- agarose
- ice
- 10× Tris-acetate-EDTA (TAE) buffer
- microcentrifuge
- balance and weigh paper
- ethidium bromide solution
- UV transilluminator
- Polaroid camera and black-and-white type 667 film
- microwave oven or hot plate

Students will need to prepare

- agarose gel
- 1× TAE

Materials per pair of students

- 1 µg of pGEM-3Zf+ cloning vector (Promega)
- 2 µg of pHGH107-containing insert of hGH (American Type Culture Collection)
- horizontal gel electrophoresis apparatus
- power supply (shared by 2 to 4 groups)
- 50-ml graduated cylinder
- 500-ml graduated cylinder

1 aliquot of *Eco*RI
1 aliquot of *Bam*HI
1 aliquot of 10× buffer
1 aliquot of sterile dH_2O
1 aliquot of 10× loading dye
1 aliquot of DNA size markers

Purification of DNA from an Agarose Gel

Class equipment and supplies
UV transilluminator
Prep-A-Gene kit (Bio-Rad, Inc.)
- DNA purification binding buffer
- DNA binding matrix
- wash solution
- elution buffer

50°C water bath
microcentrifuge

Materials per pair of students
razor blade
1 aliquot of DNA purification binding buffer
1 aliquot of DNA binding matrix
1 aliquot of wash solution
1 aliquot of elution buffer

Ligation Reaction

Class equipment and supplies
T4 DNA ligase and buffer
refrigerator
vortex

Students will need to prepare
linearized plasmid from protocol 1
hGH DNA insert from protocols 2 and 3

Materials per pair of students
1 aliquot of T4 DNA ligase
1 aliquot of 10× DNA ligation buffer
sterile dH_2O

Transformation and Plating

Class equipment and supplies
42°C water bath or heat block
37°C incubator
1 M IPTG
20-mg/ml X-Gal

Materials per pair of students
1,000-ml beaker to collect biological waste
wax marking pencil
1 plate of wild-type *E. coli* JM109 grown on Luria-Bertani (LB) agar
container with ice
1 aliquot of 50 mM $CaCl_2$
inoculating loop
Bunsen burner
2 LB agar plates
2 LB agar + ampicillin plates (100-µg/ml ampicillin)
1 aliquot of sterile LB broth
500-ml beaker containing ethanol
spreading rods

Protocols

PROTOCOL 1 LINEARIZATION OF PLASMID DNA WITH *Eco*RI AND *Bam*HI

One unit of enzyme digests 1 µg of DNA in 1 hour; enzymes are tested on lambda DNA substrates; use an excess.

See exercise 7 about restriction enzymes.

Increased incubation time ensures complete digestion.

1. Digest 1 µg of pGEM-3Zf+ with *Eco*RI and *Bam*HI. Document your digestion below, using a 20-µl volume.

2. Incubate the digestion at 37°C for 75 minutes.

3. Heat-inactivate the enzymes by incubation at 80°C for 20 minutes.

Heat denatures the enzymes so they will not interfere with subsequent ligation.

Some restriction enzymes are not inactivated by heating.

4. Determine the volume of the digestion that contains 300 ng of plasmid. Document your calculations below.

5. Remove 300 ng of digested plasmid to a clean microcentrifuge tube. Adjust the volume to 9 µl and add 1 µl of 10× loading dye.

6. Pop spin to mix.

7. Electrophorese 300 ng of linearized pGEM-3Zf+ DNA on an agarose gel (protocol 4) to check the efficiency of the digestion.

Undigested plasmids will not accept insert DNA but will transform bacteria.

8. Save the remaining linearized vector for the ligation reaction and transformation control.

PROTOCOL 2 RESTRICTION ENZYME DIGESTION OF INSERT DNA

1. Digest 2 µg of pHGH107 with *Eco*RI and *Bam*HI. Document your digestion below, using a 20-µl volume.

See exercise 7 about restriction enzymes.

2. Incubate the digestion at 37°C for 75 minutes.

Increased incubation time ensures complete digestion.

3. When the incubation is completed, add 2 µl of 10× loading dye.

Prepare gel and 1× TAE while digests are incubating.

4. Pop spin to mix.

5. Electrophorese all of the pHGH107 digestion to separate the insert DNA from the vector. The insert DNA will be gel purified in protocol 5.

Gel electrophoresis will separate enzymes from DNA, so inactivation is not required.

PROTOCOL 3 PREPARATION OF AGAROSE GEL AND 1× TAE

1. Prepare 350 ml of 1× TAE from 10× TAE stock. Document your calculations below.

TBE can be used, but DNA fragments are isolated more efficiently from gels cast and electrophoresed using TAE.

2. Prepare 50 ml of 1.0% agarose using 1× TAE and pour the gel. Document your protocol below.

See exercise 7 for assistance with gel preparation.

3. Assemble the agarose gel, electrophoresis chamber, and 1× TAE running buffer.

PROTOCOL 4 AGAROSE GEL ELECTROPHORESIS

See exercise 7 for assistance with gel electrophoresis.

DNA size markers should flank the DNA fragments of interest.

1. After adding loading dye to the samples from protocols 1 and 2, load the agarose gel. Make sure to prepare and load an appropriate DNA marker. Correlate and record sample identification with well number.

Using TAE requires a lower voltage; heat is generated and gels may melt.

2. Electrophorese the agarose gel at 100 volts for about 40 minutes.

Ethidium bromide is a mutagen and suspected carcinogen.

3. **Wearing gloves**, submerge the agarose gel in 1-µg/ml ethidium bromide solution and incubate for 5 to 10 minutes.

UV light excites the ethidium bromide, which emits light of visible wavelength.

UV light can cause skin and eye burns.

4. **Wearing gloves and using a UV face shield**, slide the agarose gel onto the UV transilluminator. Turn the transilluminator on and photograph the gel with a Polaroid camera and black-and-white film. Use an orange filter. Approximate settings are an f-stop of 8 with a 1/2-second exposure.

5. Observe the gel for DNA fragments. Linear pGEM-3Zf+ is 3,199 base pairs (bp). Digesting pHGH107 with *Bam*HI and *Eco*RI will result in a 700-bp fragment containing the hGH gene and about a 4,000-bp vector fragment.

6. The hGH gene fragment will be gel purified in protocol 5.

PROTOCOL 5 PURIFICATION OF DNA FROM AN AGAROSE GEL

UV light can cause breaks in the DNA.

The gel or gel slice may be stored at 4°C wrapped in plastic wrap.

1. **Wearing gloves**, use a razor blade to carefully chop the 700-bp hGH DNA band out of the agarose gel. You will be working on the UV transilluminator, so work quickly and protect your face and eyes from UV light exposure. *Be careful that the transilluminator is not scratched!*

2. Chop the gel slice into smaller pieces and place it into a microcentrifuge tube.

Exercises 14 and 29 use a matrix system to isolate DNA fragments; all of these systems work well.

3. Microcentrifuge the sample at 14,000 rpm for 30 seconds to force the gel to the bottom of the tube. Estimate the volume of gel.

Binding buffer dissolves agarose.

4. Add 3 volumes of DNA purification binding buffer. Document your calculations.

5. Incubate at 50°C for 15 minutes with occasional mixing. Make sure the agarose gel has completely dissolved before proceeding to the next step.

The matrix will bind DNA under high-salt conditions.

6. Resuspend the tube of matrix. Add 5 µl of the matrix suspension to the dissolved gel slice and incubate at room temperature for 10 minutes. Mix frequently by flicking the tube with your finger.

7. Microcentrifuge the tube for 30 seconds to pellet the matrix with the DNA insert bound to it.

8. Use a micropipette to remove and discard the supernatant.

9. Wash the matrix by resuspending it in 125 μl of binding buffer.

Washing removes contaminants.

10. Microcentrifuge for 30 seconds and discard the supernatant.

11. Wash the matrix two times by resuspending it in 125 μl of wash solution and then microcentrifuging for 30 seconds. Discard the supernatant each time.

12. After the final wash, microcentrifuge the matrix a second time for 30 seconds to completely remove all the liquid. Use a micropipette.

13. Add 5 μl of elution buffer. Elute the DNA from the matrix by incubation at 50°C for 10 minutes.

Low-salt conditions release DNA from the matrix.

14. Microcentrifuge at 14,000 rpm for 30 seconds to pellet the matrix.

15. Carefully collect the supernatant, containing the DNA insert, and pipette it into a clean, labeled microcentrifuge tube.

16. Calculate the concentration of insert DNA if 100% is recovered from the gel slice. The proportion of insert fragment length is proportionate to DNA quantity.

PROTOCOL 6 THE LIGATION REACTION

1. Prepare the ligation reaction.

   1 μl of linearized pGEM-3Zf+ (50 ng)
   5 μl of insert DNA (hGH)
   1 μl of 10× DNA ligation buffer
   2 μl of T4 DNA ligase
   1 μl of dH_2O
   10-μl total

2. Incubate overnight in the refrigerator.

Quantity of insert DNA is estimated to be 200 to 300 ng.

The actual quantity of insert DNA may be assessed by gel electrophoresis of 1 μl. See exercise 12.

PROTOCOL 7 DECIDING HOW TO DETECT A SUCCESSFUL LIGATION REACTION WITH BACTERIAL TRANSFORMATION

To analyze the success of the DNA ligation, the ligation reaction will be transformed into bacteria, and the bacteria will be plated and grown. You will need four plates:

- an experimental plate
- an experimental control plate
- a positive control plate
- a negative control plate

What is the control for your ligation reaction?

Use the following diagram, representing agar plates to plan plating your experiment. Label each plate with the type of medium and components it will contain and the type of transformed bacteria and ligation solution that you will spread on each. Discuss your plan with an instructor before proceeding.

PROTOCOL 8 BACTERIAL TRANSFORMATION AND PLATING

A. Preparation of Competent Bacterial Cells

Keep tubes on ice. Cold $CaCl_2$ is important for bacterial cell competence.

1. Add 250 μl of 50 mM $CaCl_2$ to two tubes. Place the tubes on ice. Label tubes with group number or initials.

2. Use a sterile inoculating loop to aseptically transfer a single colony of bacteria from the starter plate to each tube. Mix vigorously with the inoculating loop and return to ice.

Proper resuspension of bacterial cells ensures exposure to plasmids in solution.

3. Immediately place the two tubes into the vortex shaker for a few seconds to break up the clumps of bacterial cells. (Carry the tubes on ice to the vortex.) If a vortex is not available, flick the tubes vigorously with your finger. All of the clumps of bacteria should be broken up.

4. Add 10 μl of the ligation solution directly into the cell suspension in one of your tubes. Label the tube. Tap the tube directly with your finger to mix. Avoid making bubbles.

5. Add 10 μl of the control, linearized plasmid to the cell suspension in the second tube and label it. Tap the tube directly with your finger to mix. Avoid making bubbles.

6. Return both tubes to ice and incubate for 15 minutes.

B. Heat Shock of Bacterial Cells

Heat shock helps the cells to take up the plasmids.

Timing is critical.

1. Carry your tubes on ice to the water bath. Remove the tubes from ice and *immediately* immerse them in the 42°C water bath for 90 seconds. It is *critical* that cells receive a sharp and distinct shock.

2. *Immediately* return both tubes to ice and incubate on ice for 1 additional minute.

3. Place the tubes in a test tube rack at room temperature.

4. Add 250 µl of sterile liquid LB medium to each tube. Gently tap the tubes to mix.

5. Allow the bacteria to recover from the heat shock by incubation at 37°C for about 30 minutes.

During recovery, bacterial cells repair their membranes, grow, and express the genes acquired with the new plasmid DNA.

C. Plating the Transformation

1. Spread 40 µl of 20-mg/ml X-Gal onto each agar plate about 60 minutes before use.

Use proper aseptic technique when inoculating the plates.

Plates with X-Gal may be stored at 4°C.

2. Using the plan you devised in protocol 7 to detect transformation, pipette 100 µl of cells onto each plate. Label the plates.

3. Add 4 µl of 1 M IPTG to the 100 µl of bacterial cells on the agar. Spread both together onto the plate.

4. Using a sterile cell spreader, spread the bacteria over the surface of the agar in each plate. Move the spreader back and forth on the agar while turning the plate. This will spread the bacteria evenly across the agar surface.

5. Place the plates upside down in a 37°C incubator and allow them to grow for 24 to 48 hours. Alternatively, plates may incubate at room temperature for 2 to 3 days.

Inverting the plates prevents condensation from collecting on the medium.

6. Count and record the number of blue and white colonies on each plate.

Further Analysis of the DNA Ligation and Cloning Experiment (Optional)

1. Pick bacterial colonies and grow the cultures overnight. See exercise 4.

2. Miniprep of plasmid DNA. See exercise 8.

3. Restriction enzyme digestion of plasmids for inserts. See exercise 7.

4. Gel electrophoresis of digested plasmid DNA. See exercise 7.

References and General Reading

Ausubel, F. M., R. Brent, R. E. Kingston, D. D. Moore, J. G. Seidman, J. A. Smith, and K. Struhl (ed.). 1994. *Current Protocols in Molecular Biology.* Wiley Interscience, New York, N.Y.

Bio-Rad Laboratories. Prep-A-Gene kit product insert. Bio-Rad Laboratories, Hercules, Calif.

Goeddel, D. V., H. L. Heyneker, T. Hozumi, R. Arentzen, K. Itakura, D. G. Yansura, M. J. Ross, R. C. Miozzari, and P. H. Seeburg. 1979. Direct expression in *Escherichia coli* of a DNA sequence coding for human growth hormone. *Nature* **281**:546–548.

Martial, J. A., R. A. Hallewell, J. D. Baxter, and H. M. Goodman. 1979. Human growth hormone: complementary DNA cloning and expression in bacteria. *Science* **205**:602–606.

Roskman, W. G., and F. Rougeon. 1979. Molecular cloning and nucleotide sequence of the human growth hormone structural gene. *Nucleic Acids Res.* **7**:305–320.

Discussion

1. Why is the DNA fragment of interest isolated from other DNA?
2. Why is the linearized plasmid DNA heated after the restriction enzyme digestion?
3. What is the purpose of the matrix in the DNA purification procedure? Compare the purpose of this matrix with the InstaGene bead matrix used in exercise 10.
4. Consider the properties of DNA and design another method to isolate DNA from a gel slice.
5. What purposes do the IPTG and the X-Gal serve?
6. Why are ligation reaction products transformed into bacteria?
7. Why is linearized DNA transformed as a negative control?
8. Use the following diagram and draw your results. Make sure you label the type of medium in each plate and the type of bacteria you spread on each.

9. Describe the growth on your experimental plate. Is this what you expected to observe?
10. Describe the growth on your experimental control plate. Is this what you expected to observe?
11. Describe the growth on your positive control plate. Is this what you expected to observe?
12. Describe the growth on your negative control plate. Is this what you expected to observe?
13. Was your DNA ligation experiment successful? Why or why not?

14. Design a set of experiments to determine whether white or blue colonies of bacteria contain plasmids with insert fragments of DNA. Be specific. Include controls, a time line, and positive and negative results.
15. Why are some bacteria white and some bacteria blue? Draw a diagram representing how this occurs.
16. If color selection for insert DNA by disruption of β-Gal were not available, how would you determine whether or not a plasmid contained an insert?

EXERCISE 14

DNA Cloning by Polymerase Chain Reaction

Key Concepts

1. Polymerase chain reaction (PCR) may be used to clone desired fragments of DNA from complex genomes.
2. *Taq* DNA polymerase frequently adds an extra adenine (A) nucleotide to the ends of newly synthesized DNA fragments.
3. Evolutionary relationships may be studied by using PCR. Human β-globin belongs to an ancient family of oxygen-carrying molecules.

Goals

By the end of this exercise, you will be able to

- discuss the applications of PCR in DNA cloning
- describe strategies for cloning fragments of DNA using PCR
- explain the evolutionary relationships within the globin gene family

Introduction

The characteristics of PCR, the rapid amplification and production of pure, relatively large quantities of specific DNA of interest from minute amounts of impure starting material, make it possible to use PCR to clone genes. The principles of PCR were discussed fully in exercise 10. However, to use PCR to locate a specific sequence of DNA, one must know a portion of the desired gene sequence or a portion of the expressed protein sequence. This information is essential for designing or selecting suitable primers.

PCR may be used to clone known genes, known genes containing possible mutations, genes for which some amino acid sequence is available, or homologous genes within or across species. Recloning of a known sequence is relatively easy; known sequence is used to design primers, the region of interest is amplified, and the sequence is confirmed. If a researcher is studying mutations within a coding region, then primers flanking the variable sequence are designed. Mutant and variable sequences are studied by comparing the DNA sequences of many individuals.

Synthesis of degenerate primers is usually required to target DNA sequences for which only amino acid sequence is available. The genetic code is degenerate, with a number of different codons specifying many of the amino acids. Often codons for a specific amino acid differ only in the third position. For example, if short sequences of each of two peptides within a protein of interest are known, two groups of oligonucleotides can be designed to prime the PCR. Each primer group would contain oligonucleotides representing every possible codon sequence that could specify the known amino acid sequence. All the messenger RNAs of a cell or tissue type would be used to synthesize complementary DNAs (cDNAs; see exercise 29 for a thorough discussion), and the resulting pool of cDNAs would serve as the PCR template. The degenerate primers would amplify only the cDNA between the regions that code for the known peptide sequence; the primer pair corresponding to the desired sequence is the only primer pair in the two groups of degenerate primers capable of flanking the desired cDNA sequence, thereby directing amplification. Other primers are effectively ignored. The resulting high-purity, high-quantity cDNA can be tagged or labeled and used to screen cDNA or genomic DNA libraries to identify clones carrying, it is hoped, the entire gene sequence for the protein.

Another approach involves using known sequence information. Degenerate primers are designed that amplify DNA sequences of related genes within a species or homologous genes among different species. A researcher might investigate similar sequences in unrelated species to find relationships or to study the evolution of gene families. In this manner, a researcher can isolate the same or a similar gene from one organism that has already been isolated from another organism.

PCR has three steps: denaturation of the template DNA, annealing of the primers to the target sequence, and elongation of the DNA fragment flanked by the primers. As long as the primers match their complements reasonably well, DNA polymerase replicates the DNA sequence located between them. Since a researcher may be seeking to isolate a homologous gene and may not know the exact sequence, the primers don't have to match the template exactly. By reducing the annealing temperature, the stringency of primer binding to its template, one can allow some mismatches of base pairing and still copy DNA between two primers. However, after amplification, the gene sequence must be confirmed to ensure that the desired sequence was, in fact, isolated.

The Globin Gene Family

The globin gene and protein family is well understood. Hemoglobin, myoglobin, and leghemoglobin all function as oxygen transport molecules. Leghemoglobin is found in plants, myoglobin carries oxygen in vertebrate muscle, and hemoglobin carries oxygen in vertebrate blood. Molecular systematics indicates that homologs within the globin family exist in all vertebrates, in some invertebrates, and in plants participating in nitrogen fixation. Leghemoglobin of plants, a primitive oxygen-binding protein, arose from the common ancestral globin gene early in the globin family evolution. In vertebrates, globin divergences closely parallel evolutionary trends measured by conventional comparisons of anatomy and physiology.

Early mutations in vertebrate globins produced separate myoglobin and hemoglobin evolutionary pathways. Hemoglobin and myoglobin probably arose from a common ancestral gene about 500 million years ago, by a pattern of gene duplication followed by tissue specialization and divergence. Later evolution within the hemoglobin subfamily gave rise to separate α- and β-globin chains. Max Perutz, who won the Nobel Prize for his studies of hemoglobin, was the first to point out, in the 1960s, that the α- and β-globin chains of hemoglobin resembled myoglobin. A comparison of vertebrate hemoglobin sequence homologies is shown in Table 14.1.

Table 14.1 Percent difference in hemoglobins of selected vertebrates compared with humans

	% Difference from human hemoglobin						
Chain	**Gorilla**	**Rhesus monkey**	**Cow**	**Pig**	**Rabbit**	**Chicken**	**Frog**
α	NA[a]	NA	12	13	18	25	NA
β	0.7	5	17	16	10	26	45

[a]NA, not applicable.

Human hemoglobin has been well characterized. Knowledge of hemoglobin structure and genetics has permitted investigation of sickle cell anemia and the thalassemias. Recombinant DNA technologies have made it possible to express fully functional human hemoglobin in bacteria, yeasts, and mammals. Transgenic mice expressing human sickle hemoglobin provide a model system for studying current and newly devised treatments for sickle cell anemia. In addition, transgenic pigs have been produced that express high quantities of human hemoglobin; separating the human from the pig molecule was not difficult. This and other genetically engineered hemoglobins are being investigated for use in safe, effective blood substitutes, since oxygen transport is usually the most important reason for a blood transfusion. Much more work needs to be done to provide a blood substitute containing recombinantly produced hemoglobin, since free hemoglobin breaks down in animals, causing kidney damage.

Cloning with PCR

Rapid production of a desired sequence of DNA by PCR easily provides ample material for cloning. The amplified DNA fragment is then ligated into a suitable vector. This may be done by designing primers with restriction enzyme sites, using a blunt-ended ligation strategy, or taking advantage of the fact that *Taq* polymerase frequently adds an extra A nucleotide to the ends of newly synthesized fragments.

Cloning with PCR can be simplified by designing a primer pair that contains one or more restriction enzyme recognition sites on the 5′ ends. For example, a primer pair might include the *Eco*RI recognition site on primer #1 and the *Bam*HI site on primer #2, as follows: 5′-GAATTC, primer #1; 5′-GGATCC, primer #2.

During PCR cycle 1, the primers will bind to the complementary sequences on the template DNA; the restriction enzyme recognition site will not bind at all. Primer extension will produce double-stranded DNA of the sequence of interest with single-stranded recognition-site DNA overhangs at either end. During cycle 2, however, the sequence of interest *and* the recognition sites will serve as template DNA. Now primer extension will produce the desired sequence of interest with the engineered recognition sites on either end. Upon completion of the PCR cycles, the amplified product is digested with the restriction enzymes and is then ready for directional cloning into a suitable vector.

Blunt-ended ligation of PCR products may seem to be an obvious way to clone these fragments into a suitable vector. Several problems are encountered. First, blunt-ended ligations are not very efficient at ligation. There are no sticky ends to enhance insert binding with vector so the vector can close back with itself. Second, successful blunt-ended ligation usually does not result in generation of restriction enzyme sites. If a researcher wishes to move the insert DNA to another vector, this could be problematic. Finally, fragments of DNA produced by PCR frequently are not blunt-ended, but instead *Taq* polymerase adds a single A nucleotide to the 3′ end of newly synthesized DNA.

Biotechnology companies have constructed PCR cloning vectors that take advantage of this aspect of *Taq* polymerase amplifications. Special vectors containing a single thymidine (T) overhang adjacent to the multiple cloning site have been developed. The single A overhang of the PCR fragment will bind to the single thymidine residue on the vector. This is sufficient for product and vector to be joined and then ligated. Restriction enzyme digestion to prepare the PCR product for ligation and cloning is unnecessary, greatly simplifying the protocol.

In this exercise, we will obtain human DNA from cheek cells and amplify a portion of the β-globin gene from chromosome 11. Following amplification, reaction components that could interfere with the ligation are removed. The purified amplification product will be cloned into pGEM-T-Easy (Fig. 14.1), a vector that contains single T overhangs at the insert site. The recombinant vector, containing the β-globin gene sequence, will be transformed into *Escherichia coli*, and successful cloning (Fig. 14.2) will be determined by blue and white selection (described in exercise 13).

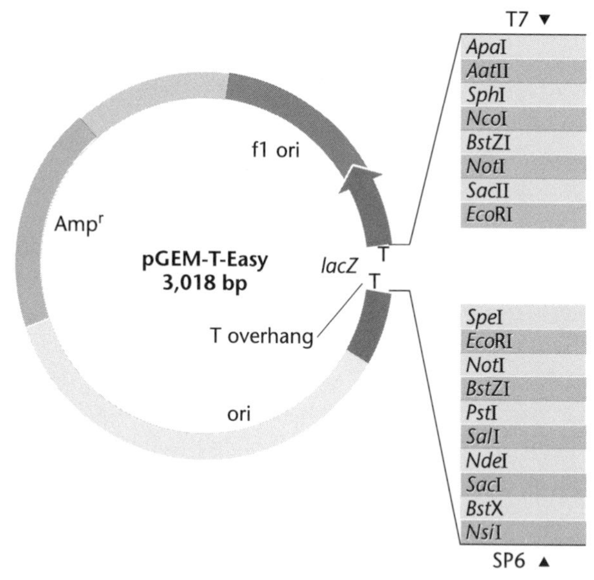

Figure 14.1 pGEM-T-Easy vector. This vector (Promega) contains single T overhangs for cloning of PCR products with A-nucleotide overhangs.

Lab Language

codon
degenerate primers
homologous gene
model organism
model system
T overhangs
transgenic

Laboratory Safety

Review the sections on physical, biological, and chemical hazards in "Laboratory Safety." Special considerations for this exercise:

hot agarose
gel electrophoresis
ethidium bromide staining
use of microbes
use of ultraviolet (UV) transilluminator
use of X-Gal (5-bromo-4-chloro-3-indolyl-β-D-galactopyranoside) and IPTG (isopropyl-β-D-thiogalactopyranoside)

Laboratory Sequence

This exercise will require four to six laboratory sessions of 3 to 4 hours each, depending on the skill level of the students, allotted class time, and extent of experimental analysis after cloning.

Day 1: Isolation of DNA from cheek cells (protocols 1 and 2)
DNA isolated in exercise 10 and stored at –20°C may be used.
Amplification of β-globin (protocol 3) (overnight)

Day 2: Agarose gel electrophoresis of amplified product (protocol 4)
Purification of amplified product (protocol 5)
Ligation reaction (protocol 6) (overnight)

Day 3: Bacterial transformation and plating (protocols 7 and 8)

Day 4: Obtain and analyze results.
Pick colonies for further analysis (optional).

Days 5 and 6: Miniprep and analysis of isolated plasmids (optional)

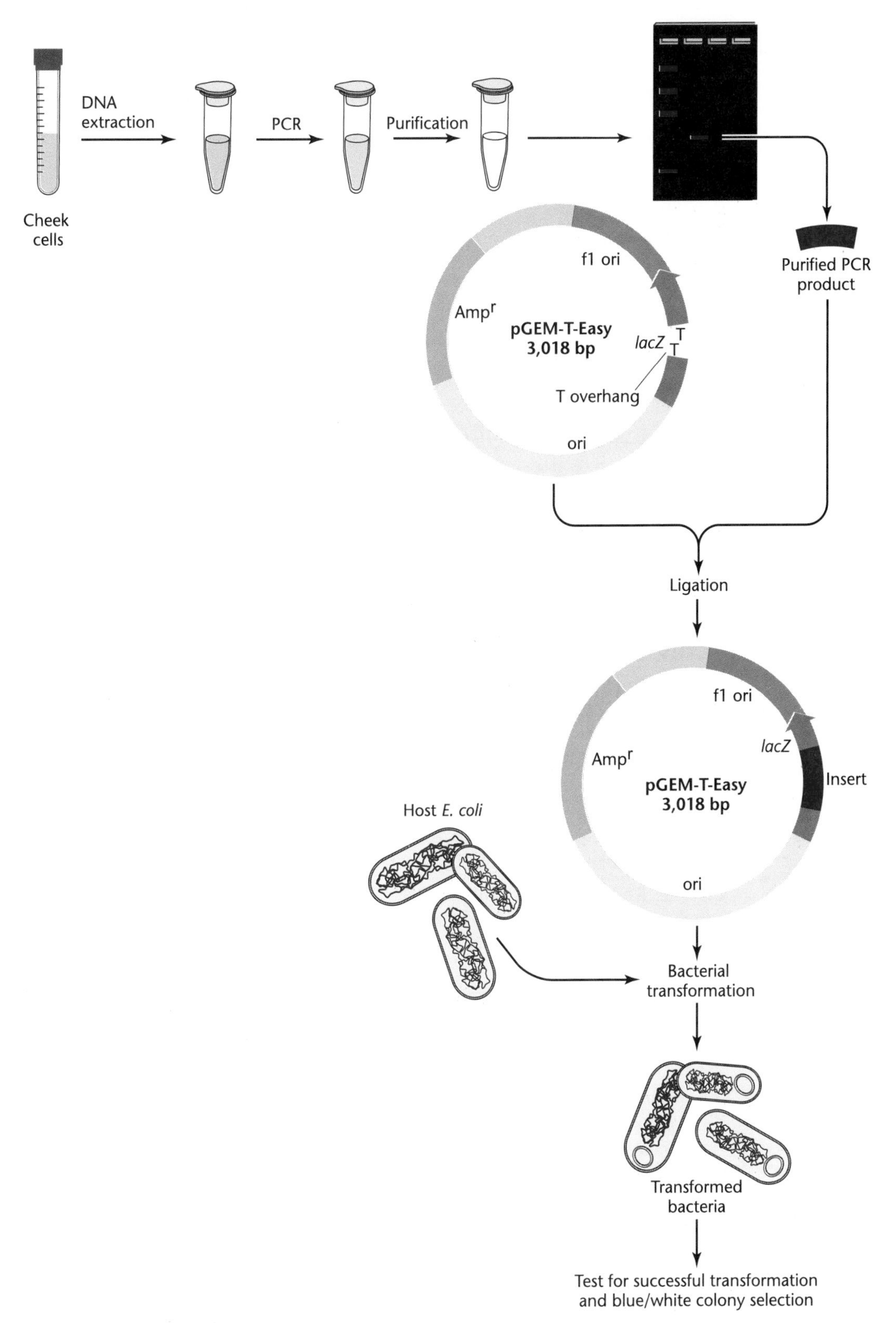

Figure 14.2 Summary of PCR cloning.

Materials

Standard Materials for the Exercise

0.5- to 10-μl micropipette and sterile tips
10- to 100-μl micropipette and sterile tips
100- to 1,000-μl micropipette and sterile tips
microcentrifuge
sterile microcentrifuge tubes
microcentrifuge test tube rack
wax or permanent marking pen
gloves

Isolation and Amplification of DNA

Class equipment and supplies
vortex
clinical centrifuge
sterile saline solution, 0.9% NaCl (exercise 3)
56°C water bath or heat block
95°C heat block (boiling water bath/hot plate)
10× PCR buffer
Taq DNA polymerase
10 mM deoxynucleoside triphosphates
β-globin primers
thermal cycler
UV transilluminator
ethidium bromide staining solution
Polaroid camera and black-and-white type 667 film

Students will need to prepare
1× Tris-borate-EDTA (TBE)
1.0% agarose gel

Materials per pair of students
2 plastic medicine cups
2 paper cups
2 500-μl aliquots of InstaGene beads
2 15-ml conical tubes
PCR reaction tubes
1 aliquot of mineral oil
1 aliquot of loading dye
1 aliquot of PCR oil
electrophoresis chambers and apparatus power supply (for 2 to 4 groups)
50-ml graduated cylinder
500-ml graduated cylinder

Purification of Amplified PCR Products

Class equipment and supplies
UltraClean PCR Clean-up kit (MoBio Laboratories, Inc.)
- SpinBind DNA binding solution
- spin filters
- SpinClean buffer
- elution buffer

microcentrifuge

Ligation Reaction

Class equipment and supplies
pGEM-T-Easy Vector System (Promega)
- pGEM-T-Easy
- T4 DNA ligase
- 2× ligation buffer

refrigerator

Students will need to prepare
purified β-globin DNA from protocol 5

Materials per pair of students
0.1-μg pGEM-T-Easy
1 aliquot of T4 DNA ligase
1 aliquot of 2× DNA ligation buffer
sterile dH_2O

Transformation and Plating

Class equipment and supplies
42°C water bath or heat block
37°C incubator
1 M IPTG
20-mg/ml X-Gal
vortex

Materials per pair of students
1,000-ml beaker for biological waste
1 plate of wild-type *E. coli* (JM109) grown on Luria-Bertani (LB) agar
container with ice
1 aliquot of 50 mM $CaCl_2$
1 inoculating loop
Bunsen burner
2 LB agar plates
2 LB agar + ampicillin plates (100-μg/ml ampicillin)
1 aliquot of sterile LB broth
500-ml beaker containing ethanol
spreading rods

Protocols

If you saved your DNA from exercise 10 and it amplified well, you may use that DNA solution instead of reisolating DNA.

PROTOCOL 1 OBTAINING CHEEK CELLS

1. Fill a plastic medicine cup with about 10 ml of sterile 0.9% NaCl solution. Rinse your mouth with the salt solution for about 10 seconds. Expel the salt solution into a paper cup.

A brief rinse will collect the normally occurring "loose" epithelial cells from the surface of the buccal mucosa.

2. Carefully pour the salt solution, now containing your cheek cells, into a 15-ml tube. Close the tube and centrifuge for 10 minutes. *Keep your cup and label your tube!*

Centrifugation will force the cells to the bottom of the tube.

3. Pour the supernatant back into the paper cup. Be careful not to disturb the cell pellet at the bottom of the tube.
4. Discard the paper cup containing the supernatant into the appropriate waste container.

PROTOCOL 2 DNA ISOLATION USING INSTAGENE BEADS

1. Resuspend the InstaGene solution with your pipette and add all of the InstaGene beads (500 µl) to your cells in the blue-capped tube. Pipette up and down to break up the clump of cells and mix them with the InstaGene beads.

Breaking cell clumps completely exposes all cells to InstaGene reagents.

2. Transfer the InstaGene beads and your cells back to the small tube. Use a vortex shaker to mix the cells and the beads.
3. Incubate your tube at 56°C for 15 minutes.

Incubation at 56°C ensures cell lysis and gently denatures DNA from associated histones.

4. Vortex again, then heat the tube for 8 minutes at 95°C in the heating block.

Incubation at 95°C completely denatures proteins from the cell.

5. Allow the tube to cool and then centrifuge for 3 minutes.

InstaGene beads chelate divalent ions, such as Mg^{2+}.

6. Carefully remove most of the supernatant, containing your DNA, from the tube. Do not pipette any InstaGene beads or cell debris from the bottom of the tube. Transfer the liquid to a clean tube.

InstaGene bead reagents and cell debris will interfere with PCR.

7. Discard the tube containing the pellet into the appropriate waste container.
8. DNA may be stored at –20°C. Use 25 µl of this DNA for PCR analysis (or from exercise 10).

PROTOCOL 3 AMPLIFICATION OF TARGET DNA

1. Prepare the PCR mixture. Label the tube on the top of the lid.

Your instructor may prepare PCR master mix for all.

Reagent	Stock concentration	Final amount	Volume
Template		Estimate 100 ng	25 µl
Primer 1	10 pmol/µl	25 pmol	
Primer 2	10 pmol/µl	25 pmol	
PCR buffer	10×	1×	
dNTPs[a]	10 mM	200 M	
Taq	5 U/µl	2.5 U	
dH_2O			
Total volume			50 µl

[a]dNTPs, deoxynucleoside triphosphates.

EXERCISE 14

You may prepare double the reaction volume without template and then split it into two tubes for experimental and negative control tubes.

2. Prepare a negative control reaction tube. Describe what you will add instead of template DNA.

A pop spin mixes and concentrates the reagents.

3. Pop spin the tube.

Some thermal cyclers have heated lids, making mineral oil unnecessary.

4. Overlay the reaction mixture with 50 µl of mineral oil.

5. Place your reaction tube in the PCR machine.

6. Amplify the target DNA using the following reaction conditions:

cycles 1 to 25	denature	94°C for 1 minute
	anneal	55°C for 2 minutes
	elongation	72°C for 3 minutes

7. After cycling is completed, carefully remove the liquid containing the amplified product or control from beneath the oil. Pipette it into a clean, labeled microcentrifuge tube.

The amplified β-globin fragment is about 200 base pairs (bp) in length.

8. Remove 9 µl of the amplified product or control to a clean tube and add 1 µl of 10× loading dye to each. Electrophorese them in an agarose gel (protocol 4) to confirm a successful reaction and proper negative control.

PROTOCOL 4 GEL ELECTROPHORESIS

1. Prepare 350 ml of 1× TBE from 10× TBE stock. Document your calculations below.

See exercise 7 for assistance with gel preparation and electrophoresis.

2. Prepare 50 ml of 1.0% agarose and pour the gel. Document your protocol below.

3. Assemble the agarose gel, electrophoresis chamber, and 1× TBE running buffer.

Sample volumes should be equal: 10 µl. Document sample preparation.

4. Pipette the amplified product and control into separate wells in the gel. Make sure to prepare and load a DNA marker. Correlate and record sample identification with well number.

5. Electrophorese the agarose gel.

Ethidium bromide is a mutagen and suspected carcinogen.

6. **Wearing gloves**, submerge the agarose gel in 1-µg/ml ethidium bromide solution and incubate for 5 to 10 minutes.

UV light excites the ethidium bromide, which emits light of visible wavelength.

UV light can cause skin and eye burns.

7. **Wearing gloves and a UV face shield**, slide the agarose gel onto the UV transilluminator. Turn the transilluminator on and photograph the gel with a Polaroid camera and black-and-white film. Use an orange filter. Approximate settings are an f-stop of 8 with a 1/2-second exposure.

The amplified β-globin fragment is about 200 bp.

8. Estimate the quantity of amplified DNA present. See exercise 12.

PROTOCOL 5 PURIFICATION OF AMPLIFIED PCR PRODUCT

Wear gloves throughout this protocol.

Exercises 13 and 29 use a matrix system to isolate DNA fragments; all work well.

1. Add 250 µl of SpinBind DNA binding solution to the remaining PCR product.

The matrix binds DNA under high-salt conditions.

2. Mix by flicking the tube with your fingers and pop spin to concentrate the reagents.

Guanidine in the binding solution denatures proteins.

3. Place a spin filter into a clean microcentrifuge tube.
4. Pipette the mixture from step 2 into the spin filter.
5. Centrifuge the tube at 14,000 rpm for 30 seconds to pellet the matrix with the DNA bound to it and to remove the liquid.
6. Remove the spin filter from the microcentrifuge tube and discard the liquid flowthrough.
7. Return the spin filter to the same microcentrifuge tube.
8. Add 100 µl of SpinClean buffer to the spin filter containing matrix and bound PCR product.

Washing removes contaminants.

9. Centrifuge the tube at 14,000 rpm for 30 seconds to pellet the matrix with the DNA bound to it and to remove the liquid.
10. Place the spin filter containing the DNA-bond matrix into a *clean* microcentrifuge tube.
11. Add 50 µl of elution buffer.

Low-salt conditions release DNA from the matrix.

12. Microcentrifuge at 14,000 rpm for 60 seconds to collect the eluted DNA. This DNA will be used for ligation with pGEM-T-Easy in protocol 6.

PROTOCOL 6 THE LIGATION REACTION

1. Prepare the ligation reaction.

It is estimated that the insert DNA will be 0.3 to 0.4 µg.

2 µl of pGEM-T-Easy (0.05 µg/µl)
4 µl of PCR-amplified insert DNA
10 µl of 2× ligation buffer
2 µl of T4 DNA ligase
2 µl of dH_2O
20-µl total

2. Incubate overnight at 4°C.

PROTOCOL 7 DECIDING HOW TO DETECT A SUCCESSFUL LIGATION REACTION WITH BACTERIAL TRANSFORMATION

To analyze the success of the DNA ligation, the ligation reaction will be transformed into bacteria, and the bacteria will be plated and grown. You will need four plates:

- an experimental plate
- an experimental control plate
- a positive control plate
- a negative control plate

What is the control for your ligation reaction?

Use the following diagram, representing agar plates, to plan plating your experiment.

Label each plate with the type of medium and components it will contain and the type of transformed bacteria that you will spread on each. Discuss your plan with an instructor before proceeding.

PROTOCOL 8 BACTERIAL TRANSFORMATION AND PLATING

A. Preparation of Competent Bacterial Cells

Keep tubes on ice. Cold $CaCl_2$ is important for bacterial cell competence.

1. Add 250 µl of 50 mM $CaCl_2$ to two tubes. Place the tubes on ice. Label the tubes with group number or initials.

2. Use a sterile inoculating loop to aseptically transfer a single colony of bacteria from the starter plate to each tube. Mix vigorously with the inoculating loop and return to ice.

Proper resuspension of bacterial cells ensures exposure to plasmids in solution.

3. Immediately place the two tubes into the vortex shaker for a few seconds to break up the clumps of bacterial cells. (Carry the tubes on ice to the vortex.) If a vortex is not available, flick the tubes vigorously with your finger. All of the clumps of bacteria should be broken up.

4. Add 10 µl of the ligation solution directly into the cell suspension in one of your tubes. Label the tube. Tap the tube directly with your finger to mix. Avoid making bubbles.

5. Add 10 µl (50 ng) of the control, linear, unligated pGEM-T-Easy vector into the cell suspension of the second tube and label it. Tap the tube directly with your finger to mix. Avoid making bubbles.

6. Return both tubes to ice and incubate for 15 minutes.

B. Heat Shock of Bacterial Cells

1. Carry your tubes on ice to the water bath. Remove the tubes from ice and *immediately* immerse them in the 42°C water bath for 90 seconds. It is *critical* that cells receive a sharp and distinct shock.

Heat shock helps the cells to take up the plasmids.

2. *Immediately* return both tubes to ice and let them stand on ice for 1 additional minute.

Timing is critical.

3. Place the tubes in a test tube rack at room temperature.

4. Add 250 μl of sterile liquid LB medium to each tube. Gently tap the tubes to mix.

5. Allow the bacteria to recover from the heat shock at room temperature for about 30 minutes.

During recovery, bacterial cells repair their membranes, grow, and express the genes acquired with the new plasmid DNA.

C. Plating the Transformation

1. Spread 40 μl of 20-mg/ml X-Gal onto each agar plate about 60 minutes before use.

Use proper aseptic technique when inoculating the plates.

X-Gal plates may be stored at 4°C.

2. Using the plan you devised in protocol 7 to detect transformation, pipette 100 μl of cells onto each plate.

3. Add 4 μl of 1 M IPTG to the 100 μl of bacterial cells on the agar. Spread both together onto the plate.

4. Using a sterile cell spreader, spread the bacteria over the surface of the agar in each plate. Move the spreader back and forth on the agar while turning the plate. This will spread the bacteria evenly across the agar surface.

5. Place the plates upside down in a 37°C incubator, and allow them to grow for 24 to 48 hours. Alternatively, plates may be incubated at room temperature for 2 to 3 days.

Inverting the plates prevents condensation from collecting on the medium.

6. Count and record the number of blue colonies and white colonies on each plate.

T-nucleotide overhangs are occasionally lost from the vector.

Further Analysis of the DNA Ligation and Cloning Experiment (Optional)

1. Pick bacterial colonies and grow the cultures overnight. See exercise 4.

2. Miniprep of plasmid DNA. See exercise 8.

3. Restriction enzyme digestion of plasmids for inserts. See exercise 7.

4. Gel electrophoresis of digested plasmid DNA. See exercise 7.

References and General Reading

Amiconi, G., and M. Brunori. 1995. Hemoglobin. *In* R. A. Meyers (ed.), *Molecular Biology and Biotechnology: a Comprehensive Desk Reference.* VCH Publishers, New York, N.Y.

Ausubel, F. M., R. Brent, R. E. Kingston, D. D. Moore, J. G. Seidman, J. A. Smith, and K. Struhl (ed.). 1994. *Current Protocols in Molecular Biology.* Wiley Interscience, New York, N.Y.

Campbell, N. A., J. B. Reese, and L. G. Mitchell. 1999. *Biology,* 5th ed. Benjamin/Cummings, Menlo Park, Calif.

Clark, D. P., and L. D. Russell. 1997. *Molecular Biology Made Simple and Fun.* Cache River Press, Vienna, Ill.

Glick, B. R., and J. J. Pasternak. 1998. *Molecular Biotechnology: Principles and Applications of Recombinant DNA,* 2nd ed. ASM Press, Washington, D.C.

Jessen, T.-H. 1995. Genetic engineering of hemoglobin. *In* R. A. Meyers (ed.), *Molecular Biology and Biotechnology: a Comprehensive Desk Reference.* VCH Publishers, New York, N.Y.

Lodish, H., D. Baltimore, A. Berk, S. L. Zipursky, P. Matsudaira, and J. Darnell. 1995. *Molecular Cell Biology,* 3rd ed. W. H. Freeman & Co., New York, N.Y.

MoBio Laboratories. 1999. *MoBio UltraClean PCR Clean-up Kit Technical Instructions.* MoBio Laboratories, Inc., Solana Beach, Calif.

Promega. 1998. *pGEM-T and pGEM-T Easy Vector Systems Technical Manual.* Promega, Madison, Wis.

Saiki, R. K., S. Scharf, F. Faloona, K. B. Mullis, G. T. Horn, H. A. Erlich, and N. Arnheim. 1985. Enzymatic amplification of beta-globin genomic sequences and restriction site analysis for diagnosis of sickle cell anemia. *Science* **230:**1350–1354.

Saiki, R. K., S. Scharf, F. Faloona, K. B. Mullis, G. T. Horn, H. A. Erlich, and N. Arnheim. 1992. Enzymatic amplification of beta-globin genomic sequences and restriction site analysis for diagnosis of sickle cell anemia—1985. *Bio/Technology* **24:**476–480.

Watson, J. D., M. Gilman, J. Witkowski, and M. Zoller. 1992. *Recombinant DNA,* 2nd ed. W. H. Freeman & Co., New York, N.Y.

Wu, D. Y., L. Ugozzoli, B. K. Pal, and R. B. Wallace. 1989. Allele-specific enzymatic amplification of beta-globin genomic DNA for diagnosis of sickle cell anemia. *Proc. Natl. Acad. Sci. USA* **86:**2757–2760.

Discussion

1. Why is PCR useful in cloning projects?
2. Describe how PCR can be used to clone nonidentical genes across species.
3. How and why is PCR used to engineer restriction enzyme sites onto cloned fragments of DNA?
4. Why was the insert DNA purified using a DNA binding matrix before cloning?
5. Why are model systems used in biology? List a specific example.
6. Discuss possible reasons for the data shown in Table 14.1.
7. Using amino acid sequence data, how might a researcher isolate the same gene from another organism? Be specific and discuss strategies for determining DNA sequence.
8. Discuss blunt-ended cloning using PCR.
9. What purposes do the IPTG and the X-Gal serve?
10. Why are ligation reaction products transformed into bacteria?
11. Why is linearized DNA transformed as a negative control?
12. Use the following diagrams and draw your results. Make sure you label the type of medium in each plate and the type of bacteria you spread on each.

13. Describe the growth on your experimental plate. Is this what you expected to observe?
14. Describe the growth on your experimental control plate. Is this what you expected to observe?
15. Describe the growth on your positive control plate. Is this what you expected to observe?
16. Describe the growth on your negative control plate. Is this what you expected to observe?
17. Was your DNA ligation experiment successful? Why or why not?
18. Design a set of experiments to determine whether white or blue colonies of bacteria contain plasmids with insert fragments of DNA. Be specific. Include controls, a time line, and positive and negative results.
19. Why are some bacteria white and some bacteria blue? Draw a diagram representing how this occurs.
20. If color selection for insert DNA by disruption of β-Gal were not available, how would you determine whether or not a plasmid contained an insert?

EXERCISE 15

DNA Cloning for Protein Expression

Key Concepts

1. Bacteria are used to produce heterologous proteins (i.e., foreign proteins, those from another species) that can be used therapeutically or to study protein structure and function.
2. Special expression vectors have been developed to allow bacteria to express a foreign protein. The vectors contain a regulatable transcription promoter, translation initiation sequences, and transcription and translation termination sequences.
3. Some mammalian proteins require posttranslational modifications for proper protein functioning that bacteria cannot carry out.
4. Other systems, such as yeast, insect cell, or Chinese hamster ovary cell culture, may be used to produce proteins that require posttranslational modifications.

Goals

By the end of this exercise, you will be able to

- explain important features of vectors used for protein expression
- select an appropriate vector for in-frame cloning of insert DNA
- understand limitations of eukaryotic protein expression in *Escherichia coli*
- discuss uses of recombinantly produced proteins

Introduction

DNA cloning, as performed in exercises 13 and 14, provides a mechanism for producing large quantities of insert DNA, which may be used for DNA sequencing or as a probe in Northern or Southern blotting. A gene's ultimate role, however, is protein production. Understanding primary DNA sequence is only one part of elucidating protein function. Gene sequences, cloned into special protein expression vectors, may be used to produce the actual protein for study or therapeutic use. Biotechnologists may even mutate DNA sequences, recombinantly produce an altered form of the protein, and study the function of particular domains of the protein.

Vectors for Protein Expression

For recombinant production of a protein, the DNA gene sequence is cloned into special plasmids called expression vectors, which have been specifically designed to produce or express proteins. They contain engineered elements necessary to produce desired proteins in large quantities (Fig. 15.1). These elements usually include

- an origin of replication
- a selectable marker, such as antibiotic resistance, to ensure maintenance of the plasmid in the microorganism

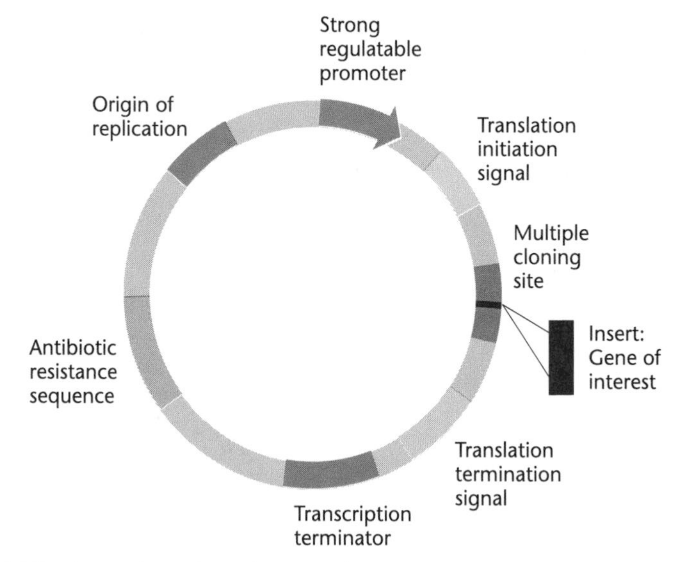

Figure 15.1 A generic prokaryotic expression vector.

- a multiple cloning site, downstream from the promoter sequence, for proper insertion and orientation of the gene fragment in the vector
- a regulatable transcription promoter, which can be induced to produce large amounts of messenger RNA from the desired gene fragment
- translation initiation sequences, which provide a ribosome binding site upstream from the initiating ATG site of the inserted gene
- transcription and translation terminators to ensure correct ending sequences for the gene

Expression of Fusion Proteins

While it seems quite easy to enable bacteria to produce large amounts of a desired protein when a promoter and start codon are present, expression of various foreign proteins in *E. coli* differs. Frequently, low levels of foreign protein expression occur in the bacterial host. To avoid some of these problems, researchers have developed fusion protein vectors. These vectors contain a short carrier DNA sequence from a gene native to the *E. coli* host. The carrier sequence encodes signals that enable protein production of most sequences ligated in the correct reading frame to its 3′ end. The N-terminal portion of the subsequently expressed protein will contain a short carrier amino acid sequence coded by the carrier DNA; hence, a fusion protein is expressed. Production of fusion proteins may reduce problems of protein expression for each new gene sequence studied.

Researchers have constructed fusion protein vectors that permit use of the N-terminal carrier protein for protein purification. By binding the fusion protein mixture from a bacterial culture lysate to the ligand of the carrier protein, the fusion protein can be separated from the total protein mixture in the lysate. Additionally, the fusion protein may contain a cleavage sequence that allows the researcher to clip the unwanted N-terminal portion from the fusion protein; then only the pure, recombinantly produced protein remains.

Uses of Recombinantly Produced Proteins

A number of proteins have been produced by means of recombinant DNA technology for research or for use as therapeutic agents. These include

- insulin
- human growth hormone
- gamma and alpha interferons
- interleukin-2
- HBsAg for hepatitis B immunization without the risk of causing disease

- granulocyte colony-stimulating factor to improve immune responses in cancer patients receiving chemotherapy
- many enzymes, including restriction enzymes, used in biotechnology

Production of these proteins is usually easier, more efficient, less costly, and safer than obtaining them from their natural sources.

Other Uses of Protein Expression

Recombinantly produced protein makes studying the protein's function easier. Molecular biologists may choose to study protein function by inducing mutations, thereby altering the amino acid sequence of the protein, or even by truncating the protein and observing the effect of the change. A protein can be truncated by removing the 3′ end of the gene DNA sequence with a restriction enzyme. The resulting sequence is cloned into an expression vector, and the truncated protein is produced. Activity of the mutated versus wild-type protein is assessed, which can elucidate the function of specific domains. Specific mutations can be introduced into a protein using polymerase chain reaction or other methodologies.

Additional Systems for Protein Expression

Yeast cells, insect cells, or mammalian cells are often used for expression of proteins from higher eukaryotic organisms. *E. coli* may not recognize the signals that direct protein synthesis, processing, modification, and secretion. Eukaryotic expression vectors are more complex than prokaryotic expression vectors but may include sequences that make them suitable for transformation into bacteria, where they can allow replication of large quantities of vector and insert DNA. Shuttle vectors such as these can transfer the gene of interest between a prokaryotic host cell and a eukaryotic host cell. Viruses such as baculovirus and vaccinia may serve as the vector for protein expression in insect cells and Chinese hamster ovary cells, respectively. For additional discussion of vectors used in eukaryotic systems, see exercise 20.

In this laboratory exercise, we will insert the mouse pancreatic amylase gene into suitable expression vectors, pGEMEX-1 and pGEMEX-2, transform *E. coli* JM109 (DE3) with the vectors, and test for expression of amylase using the standard iodine–potassium iodide method.

Starch and Amylase

Starch is a complex carbohydrate produced in algae and plants. It is a convenient energy storage molecule polymer of glucose produced during photosynthesis and an important energy source for heterotrophs. Starch molecules are polymers of a few hundred to a few thousand glucose monomers linked together via 1-4 glycosidic bonds produced by dehydration synthesis. Once polymerized, the starch molecules coil, acquiring a helical shape. A starch helix may be unbranched, as in amylose, or branched, as in amylopectin. Many such helices then aggregate to form starch granules found in roots, tubers, and seeds.

Most living organisms that are capable of using starch usually digest it extracellularly, since the molecules of starch are too large to cross the plasma membrane. Organisms using starch secrete exoamylases to accomplish digestion. Once starch molecules are cleaved into monosaccharides, membrane permeases allow cell entry. Then cellular respiration provides for glucose breakdown and ATP production. Amylase has been found in members of the five kingdoms.

The biotechnology industry uses recombinant DNA technology to produce amylases for large-scale starch processing. Large volumes of cornstarch and other grain starches are converted enzymatically into ethanol for fuel additives or into sweeteners, such as high fructose corn syrup. The global market for starch-degrading enzymes, including amylases, was reported to be $200 million in 1998. Some industrial amylases include

- SEPZYME AA 20/Maxamyl: liquid thermostable α-amylase
- Dex-lo: liquid low-temperature α-amylase

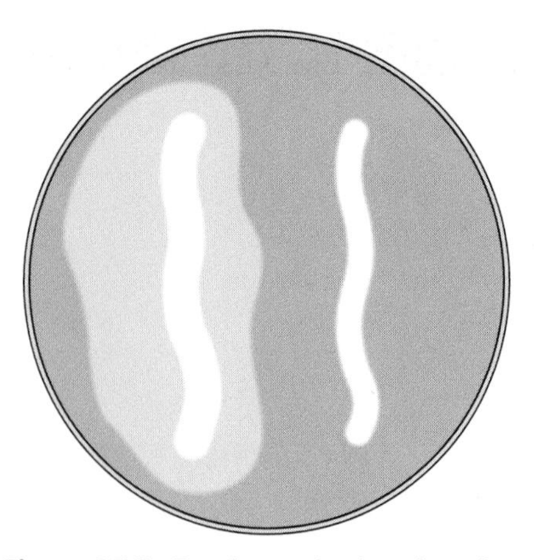

Figure 15.2 Test for production of amylase, indicating degradation of starch. Bacteria are plated onto starch agar and incubated for 48 hours. The plates are flooded with IKI solution. Amylase-positive bacteria are on the left, negative on the right.

- SEPZYME GA 300W: liquid glucoamylase for saccharification of wheat starch
- SEPZYME BBA 1500: liquid barley β-amylase for production of high and very high maltose syrups
- Mycolase: fungal α-amylase for high-conversion syrups and high maltose syrups

Additional discussion of amylases may be found in exercise 24.

The most common way to test for the presence of starch is the use of iodine–potassium iodide (IKI) solution. The iodine binds irreversibly to starch, producing a blue-black color (Fig. 15.2). If starch has been hydrolyzed, no blue-black color results, i.e., the natural yellow-brown color of the IKI will not change. This simple test reveals the presence of exoamylases.

Expression Vectors and Host Cells

In this exercise, we employ a two-step expression vector system (Fig. 15.3). The host cells, *E. coli* JM109 (DE3), have been constructed to contain the gene for T7 RNA polymerase from the phage T7, inserted downstream from the inducible *lacUV5* promoter. When the inducer, IPTG (isopropyl-β-D-thiogalactopyranoside), is provided in the growth medium, high levels of T7 RNA polymerase will be expressed. The vectors, pGEMEX-1 and pGEMEX-2, contain the phage T7 RNA polymerase promoter, ensuring transcription of insert DNA by T7 RNA polymerase. The T7 RNA polymerase promoter is situated upstream from the phage T7 gene 10 leader peptide, as is the gene 10 ribosome binding site. The position of the multiple cloning site ensures that insert DNA will be transcribed and translated as a fusion protein attached to the T7 leader peptide. This is especially useful in providing for high levels of fusion protein expression, since the T7 gene 10 fragment is abundantly expressed in *E. coli* JM109 (DE3). Often, as much as 10 mg of fusion protein per liter may be produced.

Lab Language

codon
expression
expression vector
fusion protein
in frame
reading frame
ribosome binding site
T7 DNA polymerase
T7 leader peptide
triplicate

Figure 15.3 A two-step expression vector system utilizing elements of the T7 bacteriophage.

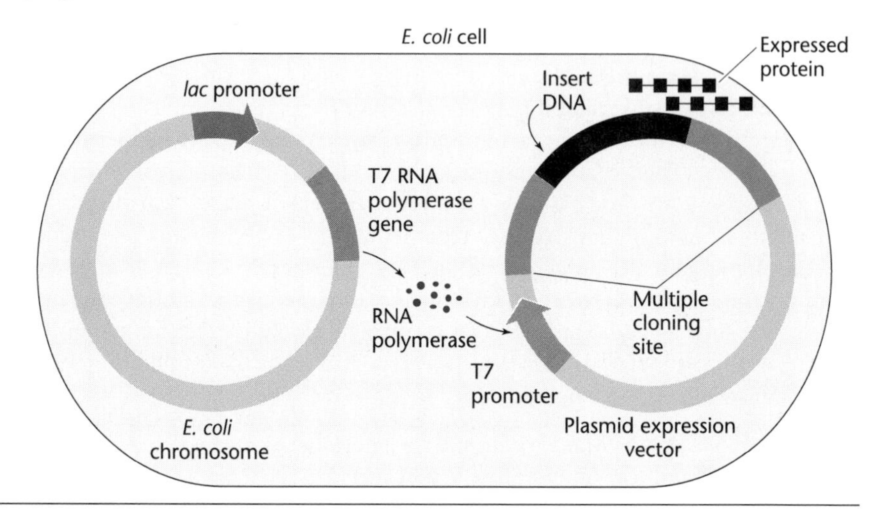

Laboratory Safety

Review the sections on physical, biological, and chemical hazards in "Laboratory Safety." Special considerations for this exercise:

hot agarose
gel electrophoresis
use of ethidium bromide
ultraviolet (UV) transilluminator
use of microbes
use of IPTG

Laboratory Sequence

This exercise will require four to six laboratory sessions of 3 to 4 hours each, depending on the skill level of the students, allotted class time, and the extent of experimental analysis after cloning.

Day 1: Selection of cloning vector (protocol 1)
This may be assigned as homework.
Restriction enzyme digestion of pGEMEX-1 and -2 and pG1-21-4 (protocols 2 and 3)
Digestions may be frozen after incubation.
Preparation of agarose gel and buffer for electrophoresis (protocol 4)
Store gel wrapped in plastic wrap at 4°C or covered with buffer at room temperature.
Gel electrophoresis (protocol 4)
Store gels or slices at 4°C.

Day 1 or 2: DNA purification (protocol 5)
Ligation reaction (protocol 6) (overnight in the refrigerator)
Streak starter culture for bacterial transformation.

Day 3: Bacterial transformation and plating (protocols 7 and 8)

Day 4: Obtain and analyze results (protocol 9).
Pick colonies for further analysis (optional).

Days 5 and 6: Miniprep and analysis of isolated plasmids (optional)

Materials

Standard Materials for the Exercise

0.5- to 10-µl micropipette and sterile tips
10- to 100-µl micropipette and sterile tips
100- to 1,000-µl micropipette and sterile tips
microcentrifuge
sterile microcentrifuge tubes
wax or permanent marking pen
gloves
microcentrifuge tube rack

Restriction Enzyme Digestion and Gel Electrophoresis

Class equipment and supplies
37°C water bath
*Hin*dIII
*Bam*HI
10× restriction enzyme buffer
agarose
10× Tris-acetate-EDTA (TAE) buffer
masking tape
dH_2O
DNA marker
ethidium bromide solution
UV transilluminator and face shields
Polaroid camera and black-and-white type 667 film
80°C water bath or heat block

Students will need to prepare
350 ml of 1× TAE
1.0% agarose gel

Materials per pair of students
gel electrophoresis chamber, gel casting tray, comb
power supply (for 2 to 4 groups)
1 aliquot of loading dye

beaker filled with crushed ice
1 aliquot of DNA size markers
1 aliquot of pGEMEX-1
1 aliquot of pGEMEX-2
1 aliquot of pG1-21-4

Keep the following reagents on ice:
1 aliquot of 10× restriction enzyme buffer
1 aliquot of *Hin*dIII
1 aliquot of *Bam*HI
1 aliquot of dH_2O

DNA Purification

Class equipment and supplies
UV transilluminator and face shields
balance
60°C water bath or heat block
E.Z.N.A. gel extraction kit (Omega Biotech)
- DNA binding buffer
- HiBind DNA spin columns
- wash buffer

1× Tris-EDTA (TE)

Materials per pair of students
razor blade
1 aliquot of binding buffer
1 spin column
1 aliquot of wash buffer
1 aliquot of 1× TE

Ligation Reaction

Class equipment and supplies
T4 DNA ligase and buffer
refrigerator

Students will need to prepare
linearized plasmids from protocol 3
amylase DNA insert from protocol 5

Materials per pair of students
1 aliquot of T4 DNA ligase
1 aliquot of 10× DNA ligation buffer
sterile dH_2O

Bacterial Transformation, Plating, and Detection of Amylase Production

Class equipment and supplies
42°C water bath or heat block
37°C incubator
1 M IPTG
IKI solution
vortex

Materials per pair of students
1,000-ml beaker for biological waste
wax marking pencil
1 plate of *E. coli* JM109 (DE3) grown on Luria-Bertani (LB) agar
container with ice
1 aliquot of 50 mM $CaCl_2$
inoculating loop
Bunsen burner
2 LB starch agar plates
2 LB starch agar + ampicillin plates
1 aliquot of sterile LB broth
500-ml beaker containing ~50-ml ethanol
spreading rods

Protocols

PROTOCOL 1 SELECTION OF THE CLONING VECTOR

1. Figure 15.4 shows the complementary DNA sequence of the gene for mouse amylase. Locate the start ATG sequence and note any unique restriction enzyme sites 5′ to the ATG start. Note restriction enzyme sites that are found internal to the gene sequence. You will digest this DNA with *Bam*HI and *Hin*dIII.

2. Figure 15.5 shows the pGEMEX-1 and pGEMEX-2 vectors and their reading frames. Select the pGEMEX vector that can be used to express mouse amylase in frame when cloned directionally using *Bam*HI and *Hin*dIII.

3. Which restriction enzyme can be used to match sticky ends from the vector and the insert DNA to allow in-frame cloning of mouse amylase?

4. Discuss your strategy with your instructor.

```
   1 ggatcctcaa agcaaaatga agttcgttct gctgctttcc ctcattgggt tctgctgggc
  61 tcaatatgac ccacatactt cagatgggag gactgctatt gtccacctgt tcgagtggcg
 121 ctgggttgat attgccaagg aatgtgagcg atacttagct cctaagggat ttggaggagt
 181 gcaggtctct ccacccaatg aaaacattgt agttcataac ccatcaagac cttggtggga
 241 aagataccaa ccaatcagct ataaaatctg tacaaggtct ggaaatgaag atgaattcag
 301 agacatggtg acaaggtgca acaatgttgg tgtccgtatt tatgtggatg ctgtcattaa
 361 ccacatgtgt ggctcaggca atcctgcagg aacaagcagt acctgtggaa gttacctcaa
 421 tccaaataac agggaattcc cagcagttcc atactctgct tgggacttta acgataataa
 481 atgtaatgga gaaattagta actacaatga tgcttatcag gtcagaaatt gtcgtgtgac
 541 tggccttctg gatcttgcac ttgagaaaga ttatgttcga accaaggtgg ctgactatat
 601 gaaccatctc attgacattg gagtagcagg gttcagactt gatgctgcta agcacatgtg
 661 gcctagagac ataaaggcag ttttggacaa attgcataat ctcaatacaa aatgattctc
 721 ccaaggaagc agacctttca ttttccaaga ggtcattgat ctgggtggtg aggcagttaa
 781 aggtagtgag tactttggaa atggccgtgt gacagaattc aagtttggtg taaaacttgg
 841 cacagttatc cgcaagtgga atggcgagaa gatgtcctat ttaaagaact ggggagaagg
 901 ttggggtatg gtgccttctg acagagccct tgtgtttgtg gacaaccatg ataatcagcg
 961 aggacatggt gctggaggat catccatcct gacattctgg gatgctagaa tgtataaaat
1021 ggctgtcgga tttatgttgg ctcatcctta tggattcaca agagtaatgt caagttaccg
1081 ttggaataga aatttccaga atggaaaaga tcagaatgac tggattggac cacccaataa
1141 caatggagta acaaaagaag tgaccattaa tgcagacact acttgtggca atgactgggt
1201 ctgtgaacat agatggcgtc aaatcaggaa catggttgcc ttcagaaatg tggtcaatgg
1261 tcagcctttt tcaaactggt gggataataa cagcaaccaa gtagctttta gcagaggaaa
1321 cagaggattc attgtcttta acaatgatga ctgggctttg tcagccactt tacagactgg
1381 tcttcctgct ggcacatact gtgatgtcat ttctggagat aaggtcgatg gcaattgcac
1441 tggacttaga gtgaatgttg gcagtgatgg caaagctcac ttttccatta gtaactctgc
1501 tgaggaccca tttattgcaa tccatgctga ctcaaaattg taagaatcta tattaaagag
1561 atttggatta agcatcaagctt
```

Figure 15.4 The nucleotide sequence of the mouse amylase gene.

PROTOCOL 2 RESTRICTION ENZYME DIGESTION OF INSERT DNA

1. Digest the plasmid pG1-21-4 containing the DNA for mouse amylase with *Bam*HI and *Eco*RI.

2. Document your restriction enzyme digestion below.

   10 μl of pG1-21-4 (0.2 μg/μl)
   2 μl of 10× restriction enzyme buffer
   2 μl of 10 U/μl *Bam*HI
   2 μl of 10 U/μl *Eco*RI
   4 μl of dH_2O
   20-μl total

Figure 15.5 The amino acid reading frames of pGEMEX-1 and pGEMEX-2. (A) The pGEMEX-1 vector DNA sequence, indicating reading frames for amino acids. (B) The pGEMEX-2 vector DNA sequence, indicating reading frames for amino acids. The two boldface G nucleotides indicate additions to the sequence, causing a shift in the reading frame.

A

5′ TTA TTA ACC CTC ACT AAA GGG AAG GCC AAG TCG GCC GAG CTC GAA TTC GTC GAC CTC GAG GGA TCC GGG CCC TCT AGA
*Sac*I *Eco*RI *Sal*I *Xho*I *Bam*HI *Apa*I

TGC GGC CGC ATG CAT AAG CTT GAG TAT TCT ATA GTG TCA CCT AAA TCC 3′
*Not*I *Nsi*I *Hind*III

B

5′ TTA TTA ACC CTC ACT AAA GGG AAG G**G**C CAA GTC GGC CGA GCT C**G**G AAT TCG TCG ACC TCG AGG GAT CCG GGC CCT CTA
*Sac*I *Eco*RI *Sal*I *Xho*I *Bam*HI *Apa*I

GAT GCG GCC GCA TGC ATA AGC TTG AGT ATT CTA TAG TGT CAC CTA AAT CC 3′
*Not*I *Nsi*I *Hind*III

EXERCISE 15

Prepare the digestion of vector DNA and incubate while this digestion is incubating.

3. Incubate the digestion at 37°C for at least 75 minutes.
4. Electrophorese the digestion to separate the insert DNA from the plasmid. The insert will be gel purified in protocol 5.

PROTOCOL 3 LINEARIZATION OF pGEMEX VECTOR DNA

1 unit of enzyme digests 1 μg of DNA in 1 hour.

See exercise 7 about restriction enzymes.

1. Digest 1 μg of pGEMEX-1 and 1 μg of pGEMEX-2 with *Bam*HI and *Eco*RI. Document your digestion below.
2. Incubate the digestion at 37°C for 75 minutes.

Heat denatures Eco*RI so it will not interfere with subsequent ligation.*

Some restriction enzymes are not inactivated by heating.

3. Heat-inactivate the enzyme by incubation at 80°C for 20 minutes.
4. Determine the volume of digest that contains 300 ng of DNA. Document your calculation below.
5. Remove 300 ng of each vector to a clean microcentrifuge tube. Adjust the volumes to 9 μl with sterile dH_2O or 1× TE and add 1 μl of 10× loading dye to each sample.
6. Pop spin to mix.
7. Electrophorese 300 ng of each linearized pGEMEX vector on an agarose gel to check the efficiency of the digestion (in protocol 4).

PROTOCOL 4 GEL ELECTROPHORESIS

TBE can be used, but DNA fragments are isolated more efficiently from gels cast and electrophoresed using TAE.

1. Prepare 350 ml of 1× TAE from 10× TAE stock. Document your calculations below.

See exercise 7 for assistance with gel preparation.

2. Prepare 50 ml of 1.0% agarose in 1× TAE and pour the gel. Document your protocol below.
3. Assemble the agarose gel, electrophoresis chamber, and 1× TAE running buffer.
4. After adding loading dye to the samples, load the agarose gel. Make sure to prepare and load a DNA marker. Correlate and record sample identification with well number.

5. Electrophorese the agarose gel.

6. **Wearing gloves,** submerge the agarose gel in 1-µg/ml ethidium bromide solution and incubate for 5 to 10 minutes.

7. **Wearing gloves and a UV face shield,** slide the agarose gel onto the UV transilluminator. Turn the transilluminator on and photograph the gel with a Polaroid camera and black-and-white film. Use an orange filter. Approximate settings are an f-stop of 8 with a 1/2-second exposure.

EXERCISE 15

Ethidium bromide is a mutagen and suspected carcinogen.

UV light excites the ethidium bromide, which emits light of visible wavelength.

UV light can cause skin and eye burns.

PROTOCOL 5 PURIFICATION OF DNA FROM AN AGAROSE GEL

1. Weigh a clean, empty microcentrifuge tube. Record the weight below.

2. **Wearing gloves,** use a razor blade to carefully chop the band of interest out of the agarose gel. You will be working on the UV transilluminator, so work quickly and protect your face and eyes from UV light exposure. *Do not scratch the surface of the transilluminator.*

UV light can cause breaks in the DNA.

Chop close to the DNA to minimize extra agarose gel.

3. Chop the gel slice into smaller pieces and place it into a microcentrifuge tube.

4. Determine the weight of the gel pieces by weighing the tube and gel pieces. Document your calculation below.

5. Determine the gel volume using 1 g = 1 ml as a conversion factor. Document your calculation below.

6. Add 4 volumes of DNA binding buffer. Document the amount added below.

Binding buffer dissolves agarose.

7. Incubate at 60°C for 7 minutes or until the gel has dissolved, with occasional mixing. Make sure the agarose gel has completely dissolved before proceeding to the next step.

8. Place a HiBind DNA spin column into a clean microcentrifuge tube.

9. Add the dissolved gel-binding buffer mixture to the spin column. Add only 800-µl volume at one time.

The matrix will bind DNA under high-salt conditions.

10. Microcentrifuge the tube at 14,000 rpm for 1 minute.

11. Discard the liquid flowthrough.

12. Repeat steps 9, 10, and 11 if necessary, until all of the mixture from step 7 has been filtered.

13. Wash the column by adding 750 µl of wash buffer.

Washing removes contaminants.

14. Microcentrifuge at 14,000 rpm for 1 minute and discard the supernatant.

15. Repeat steps 13 and 14.

Low-salt conditions release DNA from the matrix.

16. Centrifuge the column again at 14,000 rpm for 3 minutes to dry the filter.

17. Place the filter into a clean, labeled microcentrifuge tube.

18. Add 30 μl of 1× TE to the column and elute the DNA by centrifugation at 14,000 rpm for 1 minute.

PROTOCOL 6 THE LIGATION REACTION

1. Prepare two ligation reactions, with pGEMEX-1 and pGEMEX-2.

   5 μl of linearized vector (50 ng)
   10 μl of insert DNA (The quantity of insert DNA may be assessed by gel electrophoresis of 1 μl. See exercise 12.)
   2.2 μl of 10× ligation buffer
   4 μl of T4 DNA ligase
   0.8 μl of dH_2O
   22 μl total

2. Incubate overnight at 4°C.

PROTOCOL 7 DECIDING HOW TO DETECT A SUCCESSFUL LIGATION REACTION WITH BACTERIAL TRANSFORMATION

To analyze the success of the DNA ligation, transform the ligation reaction into bacteria and plate and grow the bacteria. You will need four plates:

- an experimental plate
- a positive growth control plate for each ligation reaction

Use the following diagram, representing agar plates, to plan plating your experiment.

Label each plate with the type of medium and components it will contain and the type of transformed bacteria you will spread on each. Discuss your plan with an instructor before proceeding.

PROTOCOL 8 BACTERIAL TRANSFORMATION AND PLATING

A. Preparation of Competent Bacterial Cells

1. Add 250 µl of 50 mM $CaCl_2$ to two tubes. Place the tubes on ice. Label the tubes with "1" for the pGEMEX-1 ligation and "2" for the pGEMEX-2 ligation.

Keep tubes on ice. Cold $CaCl_2$ is important for bacterial cell competence.

2. Use a sterile inoculating loop to aseptically transfer a single colony of bacteria from the starter plate to each tube. Mix vigorously with the inoculating loop and return to ice.

3. Immediately place the two tubes into the vortex shaker for a few seconds to break up the clumps of bacterial cells. (Carry the tubes on ice to the vortex.) If a vortex is not available, flick the tubes vigorously with your finger. All of the clumps of bacteria should be broken up.

Proper resuspension of bacterial cells ensures exposure to plasmids in solution.

4. Add 10 µl of the appropriate ligation solution directly into the cell suspension of each tube. Tap the tube directly with your finger to mix. Avoid making bubbles.

5. Return both tubes to ice and incubate for 15 minutes.

B. Heat Shock of Bacterial Cells

1. Carry your tubes on ice to the water bath. Remove the tubes from ice and *immediately* immerse them in the 42°C water bath for 90 seconds. It is *critical* that cells receive a sharp and distinct shock.

Heat shock helps the cells to take up the plasmids.

Timing is critical.

2. *Immediately* return both tubes to ice and let them stand on ice for 1 additional minute.

3. Place the tubes in a test tube rack at room temperature.

4. Add 250 µl of sterile liquid LB medium to each tube. Gently tap the tubes to mix.

5. Allow the bacteria to recover from the heat shock at room temperature for about 30 minutes.

During recovery, bacterial cells repair their membranes, grow, and express the genes acquired with the new plasmid DNA.

C. Plating the Transformation

1. Using the plan you devised in protocol 7 to detect transformation, pipette 100 µl of cells onto each LB starch agar plate and each LB starch agar + ampicillin plate.

Use proper aseptic technique when inoculating the plates.

2. Add 4 µl of 1 M IPTG to the 100 µl of bacterial cells on the agar. Spread both together onto the plate.

3. Using a sterile cell spreader, spread the bacteria over the surface of the agar in each plate. Move the spreader back and forth on the agar while turning the plate. This will spread the bacteria evenly across the agar surface.

4. Place the plates upside down in a 37°C incubator and allow them to incubate for 48 hours. Alternatively, plates may be incubated at room temperature for 3 to 4 days.

Inverting the plates prevents condensation from collecting on the medium.

Amylase production requires 48 hours of growth.

5. Count and record the number of colonies on each plate.

PROTOCOL 9 IDENTIFICATION OF AMYLASE-PRODUCING CLONES

1. Observe the plates.
2. **Working aseptically and wearing gloves,** flood the surface of the agar with about 5 ml of IKI solution.
3. Immediately remove the IKI solution and discard it in an appropriate container.
4. Observe the plates for colonies with a clearing around the outside.

Further Analysis of the DNA Ligation and Cloning Experiment (Optional)

1. Pick bacterial colonies and grow the cultures overnight. See exercise 4.
2. Miniprep of plasmid DNA. See exercise 8.
3. Restriction enzyme digestion of plasmids for inserts. See exercise 7.
4. Gel electrophoresis of digested plasmid DNA. See exercise 7.

References and General Reading

Ausubel, F. M., R. Brent, R. E. Kingston, D. D. Moore, J. G. Seidman, J. A. Smith, and K. Struhl (ed.). 1994. *Current Protocols in Molecular Biology.* Wiley Interscience, New York, N.Y.

Doyle, K. (ed.). 1996. *Protocols and Applications Guide,* 3rd ed. Promega Corporation, Madison, Wis.

Glick, B. R., and J. J. Pasternak. 1998. *Molecular Biotechnology: Principles and Applications of Recombinant DNA,* 2nd ed. ASM Press, Washington, D.C.

Gumucio, D. L., K. Wiebauer, R. M. Caldwell, L. C. Samuelson, and M. H. Meisler. 1988. Concerted evolution of human amylase genes. *Mol. Cell. Biol.* **8:**1197–1205.

Gumucio, D. L., K. Wiebauer, A. Dranginis, L. C. Samuelson, L. O. Treisman, R. M. Caldwell, T. K. Antonucci, and M. H. Meisler. 1985. Evolution of the amylase multigene family. YBR/Ki mice express a pancreatic amylase gene which is silent in other strains. *J. Biol. Chem.* **260:**13483–13489.

Hagenbuchle, O., O. Bovey, and R. A. Young. 1980. Tissue-specific expression of mouse-alpha-amylase genes: nucleotide sequence of isoenzyme mRNAs from pancreas and salivary gland. *Cell* **21:**179–187.

Lodish, H., D. Baltimore, A. Berk, S. L. Zipursky, P. Matsudaira, and J. Darnell. 1995. *Molecular Cell Biology,* 3rd ed. W. H. Freeman & Co., New York, N.Y.

Meisler, M., J. Strahler, K. Wiebauer, and K. K. Thomsen. 1983. Multiple genes encode mouse pancreatic amylases. *Isozymes Curr. Top. Biol. Med. Res.* **7:**39–57.

San, K.-Y., and G. N. Bennett. 1995. Expression systems for DNA processes. *In* R. A. Meyers (ed.), *Molecular Biology and Biotechnology: a Comprehensive Desk Reference.* VCH Publishers, New York, N.Y.

Schibler, U., P. H. Shaw, F. Sierra, O. Hagenbuchle, P. K. Wellauer, M. Carneiro, and R. Walter. 1986. Structural arrangement and tissue-specific expression of the two murine alpha-amylase loci Amy-1 and Amy-2. *Oxf. Surv. Eukaryot. Genes* **3:**210–234.

Studier, F. W., and B. A. Moffatt. 1986. Use of bacteriophage T7 RNA polymerase to direct selective high-level expression of cloned genes. *J. Mol. Biol.* **189:**113–134.

Watson, J. D., M. Gilman, J. Witkowski, and M. Zoller. 1992. *Recombinant DNA,* 2nd ed. W. H. Freeman & Co., New York, N.Y.

Wiebauer, K., D. L. Gumucio, J. M. Jones, R. M. Caldwell, H. T. Hartle, and M. H. Meisler. 1985. A 78-kilobase region of mouse chromosome 3 contains salivary and pancreatic amylase genes and a pseudogene. *Proc. Natl. Acad. Sci. USA* **82:**5446–5449.

Discussion

1. How do expression vectors differ from vectors used for replicating DNA?
2. What is a fusion protein expression vector?
3. Discuss potential problems in expressing mammalian proteins in *E. coli.* How are these problems resolved?
4. What are the advantages and disadvantages of recombinantly produced proteins?
5. Discuss a specific example of your own use of a recombinant protein.
6. When designing a scheme for production of a recombinant protein, what must be taken into consideration?
7. Describe and discuss an experiment that may use a recombinant protein.
8. Use the following diagram to draw your results. Make sure you label the type of medium in each plate and the type of bacteria you spread on each.

9. Describe the growth on each of your plates. Is this what you expected to observe?
10. Which vector resulted in production of amylase? Was this what you predicted?
11. Did you identify another restriction enzyme digestion scheme that would have resulted in cloning amylase in frame?

SECTION IV

Southern Blotting

EXERCISE 16

Restriction Fragment Length Polymorphism Analysis of Human DNA

Key Concepts

1. Southern blotting is the process of transferring and immobilizing DNA onto a membrane after restriction enzyme digestion and electrophoresis. The purpose is to subsequently probe the membrane with DNA to find complementary sequences.
2. A complementary fragment of DNA is used to detect specific DNA sequences immobilized onto a membrane from a Southern blot. DYZ1 is a Y chromosome-specific probe, and D2S44 detects a restriction fragment length polymorphism (RFLP) on chromosome 2.
3. Long, relatively intact pieces of DNA are required for restriction enzyme digestion for Southern blot analysis of human genomic DNA. Approximately 10- to 20-μg quantities are required to identify a single-copy region from the human genome.

Goals

By the end of this exercise, you will be able to

- understand the process of Southern blotting
- discuss the applications of Southern blotting
- describe an example of an RFLP
- discuss the uses of RFLP analysis

Introduction

Southern hybridization, also called Southern blotting, is a procedure developed in 1975 by E. M. Southern to analyze specific restriction enzyme fragments of DNA within the background of a mixture of many different restriction fragments (Fig. 16.1). The first step in Southern blotting is to cut the genomic DNA with one or more restriction enzymes, then generate a restriction enzyme fragment pattern of the DNA by agarose gel electrophoresis. Denaturation of the double-stranded DNA into single-stranded DNA follows. Next, the single-stranded DNA is transferred from the agarose gel to a nylon or nitrocellulose membrane, often by capillary action. The DNA fragments bind to the membrane in the identical pattern as in the agarose gel. The membrane becomes a replica of the gel. Because the DNA is single stranded, a region of interest can be identified by exploring the genomic DNA with a probe whose sequence is complementary to the region of interest. The probe will hybridize to the specific complementary sequence on the genomic DNA bound to the membrane. The DNA of interest is made visible by means of a probe labeled with radioactivity or with a colorimetric reagent system.

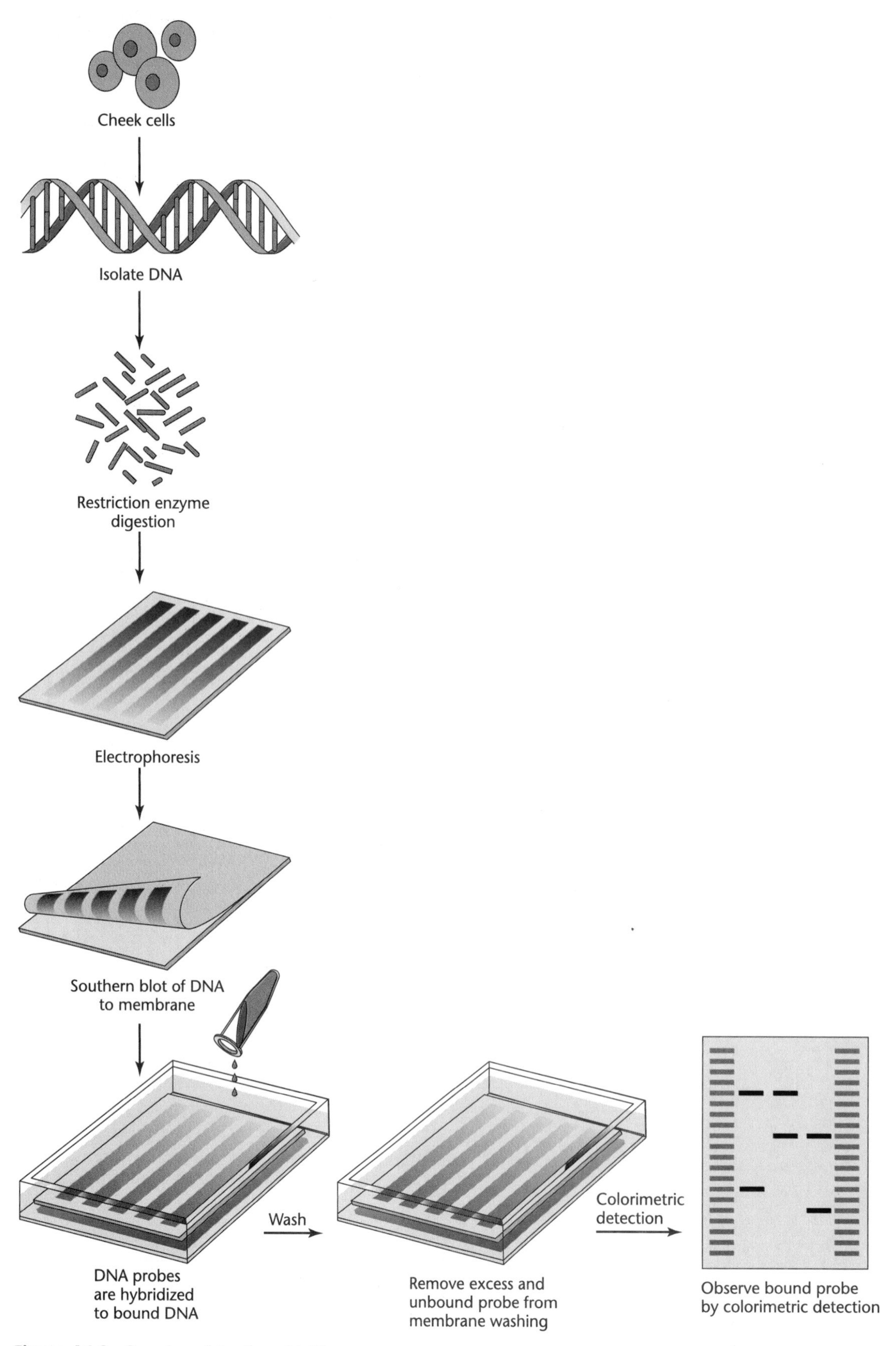

Figure 16.1 Overview of Southern blotting.

The most critical component of the Southern blot is the quantity and quality of the DNA. High-quality DNA consists of long, intact strands; the length of the strands depends on the length of the target DNA to be analyzed. If the gene or sequence of interest is only a few hundred base pairs in length, then random shearing during the genomic DNA isolation procedure is less likely to break the DNA in the middle of the desired fragment. If, however, the DNA of interest is a sequence 10,000 base pairs (bp) in length, then it may be broken in the middle, producing many restriction enzyme fragments. This results in the probe's binding to many restriction fragments of different lengths instead of binding to only one fragment.

The quantity of DNA used for a Southern blot depends on the complexity of the genome being analyzed and the sensitivity of the detection system. The more sensitive the method, the less DNA necessary for analysis. A commonly used colorimetric system, the Boehringer Mannheim Genius Kit, has a 0.1-pg sensitivity. Since the human genome contains 3 billion base pairs of DNA, 10 to 20 μg of DNA is required to identify a specific region of interest in a Southern blot. Another way to look at this is to consider that a 500-bp sequence would be equivalent to 1.5 pg of DNA. When this region is present as a single copy in the human genome, then 10 μg of total DNA is required to ensure that the specific region can be identified. Therefore, digesting, electrophoresing, and transferring 10 μg of total genomic DNA should contain a sufficient quantity of DNA to detect the gene of interest using the Genius system.

Once the restriction fragments of the DNA of interest have been separated by gel electrophoresis, the gel must be prepared for efficient transfer of the DNA. First, the gel is submerged in denaturation buffer. This step separates the double-stranded DNA into single-stranded DNA. Next, the gel is placed in a neutralization buffer. The neutralization step lowers the pH of the gel and the DNA within it so that the DNA can bind to the membrane. Finally, the DNA is transferred out of the gel and trapped onto the membrane by capillary action. Salt buffer moves through the gel toward the membrane, carrying the single-stranded DNA to the membrane where binding takes place. Capillary transfer is the most popular, inexpensive, and simple method for the transfer of DNA from agarose gels to membranes. The transfer takes about 12 to 18 hours, depending on the length of DNA fragments.

When the transfer is complete, the DNA must be irreversibly bound to the membrane. This can be performed either by baking the membrane at 80°C for 2 hours under vacuum or by ultraviolet (UV) cross-linking the DNA to the membrane. UV cross-linking is quicker and more efficient, especially if a membrane is going to be analyzed more than once.

Analysis of a Southern blot typically involves the use of a DNA probe. The DNA probe is complementary to the targeted DNA sequence. Because DNA base pairs have great specificity, a complementary DNA probe can be used to locate or "fish out" a desired DNA sequence. The conditions under which the probe is incubated with the target sequence and the conditions that are used to wash excess probe from the DNA bound to the membrane are very important to Southern blotting. Since DNA strands bind to each other by weak hydrogen bonds, conditions that promote the breaking of these bonds are considered stringent. Probes that remain bound to the target DNA under conditions of high stringency, high temperature, and low salt concentration match their complement well. Conditions of low stringency, low temperature, and high salt concentration are used to promote hydrogen bonding when the probe does not match its complement well.

Most gene-coding segments of DNA are identical from one individual to another. Gene-coding regions, however, compose only about 5% of the 3 billion base pairs of DNA that make up the human genome. This leaves most of the DNA with no known function. The noncoding regions of the genome are sometimes called junk DNA. Useful RFLPs can be found within the noncoding stretches of

DNA. An RFLP results when a restriction enzyme recognition site is either present or absent in a region of DNA. When DNA is digested by the particular restriction enzyme, the resulting fragments may be long or short, depending on whether a recognition site is present in that region of the DNA. In addition to noncoding regions of DNA, the sequences coding for blood group antigens and the major histocompatibility complex (MHC) antigens, MHC-I and MHC-II, are also highly variable in the human population.

Sometimes these differences in fragment lengths are part of the normal variation in DNA that exists from person to person. Because these individual differences in RFLPs exist and are inherited, they can be used to identify criminal suspects or to determine parentage. Other times, RFLPs may be associated with genetic disorders and can be used to diagnose genetic diseases or inherited disorders prenatally. To include or exclude parents or to identify criminal suspects, specific regions of DNA are examined that have been shown to be highly polymorphic within the human population.

Two probes, DYZ1 and D2S44, are commonly used for identification of individuals. DYZ1 is a sequence of DNA found on the Y chromosome; it will bind only to DNA samples isolated from males. D2S44 is a region of DNA found on chromosome 2. There are over 70 alleles of D2S44, making it highly polymorphic. It has been well characterized in various ethnic populations, so it can be used to identify individuals.

The probe serves as a label or tag to identify specific regions of DNA out of the total 3 billion base pairs of DNA found in the human genome. For visualization of the probe hybridized to its complement, it must be labeled with a reagent to make it visible. One of the first labels coupled to a probe was radioactive phosphorus, ^{32}P, which is incorporated directly into the probe DNA. The probe is detected by autoradiography. The membrane is compressed against X-ray film, and the radioactive phosphorus exposes the film only in the positions that mirror the fragments to which the probe has bound. Radiolabeled probes provide a very sensitive method for detecting complementary DNA, but the probes are usable for only about a week or two owing to the short 14-day half-life of ^{32}P. In addition, those working with radioactive elements must be licensed, and the probes pose hazards in the laboratory and in disposal.

Biotechnology companies have developed sensitive, nonradioactive colorimetric methods for probe labeling and detection. One such assay system is the digoxigenin (DIG) labeling method from Boehringer Mannheim Biochemicals. Digoxigenin is a highly antigenic plant steroid; antibodies against it will not cross-react with antigens from other organisms. The Boehringer Mannheim method incorporates uracil, coupled to a digoxigenin molecule, into the probe DNA. To detect whether the probe has hybridized to its complement, one must use an antibody specific to digoxigenin. The antibody is tagged with a reagent, which causes a visible colorimetric reaction when it binds to the digoxigenin. If the reagent is not activated and no color develops, it means that no antibody has bound to digoxigenin; therefore, no probe has bound to its complement (Fig. 16.2 and 16.3).

The colorimetric reagent system consists of the antidigoxigenin antibody coupled to the enzyme alkaline phosphatase (AP), 5-bromo-4-chloro-3-indolyl-phosphate (BCIP), and nitroblue tetrazolium (NBT). AP hydrolyzes phosphate moieties from the BCIP, which then oxidizes the NBT, resulting in a blue-purple precipitate.

Processes similar to Southern blotting have been developed to examine messenger RNA and proteins. They are humorously but appropriately called Northern blotting for the transfer of messenger RNA and Western blotting for the transfer of proteins. A Northern blot is probed with complementary DNA, and a Western blot is probed with antibody specific for the protein of interest.

Probe binding to its complement

DNA sequence of interest

DNA ACGTTCCGCTTAACGATAGC

1 Hybridization of target DNA with probe

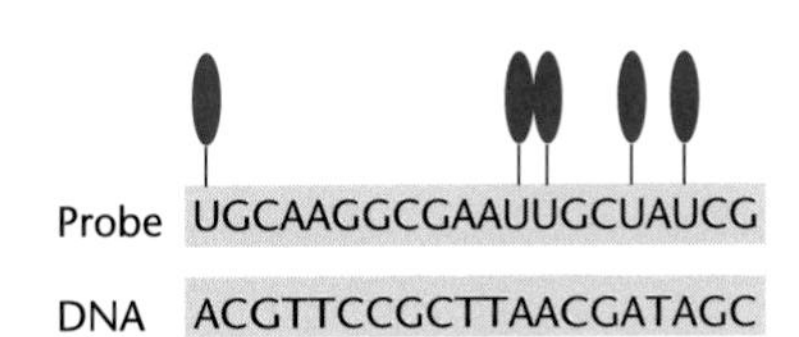

2 Unbound, excess probe is removed from target DNA

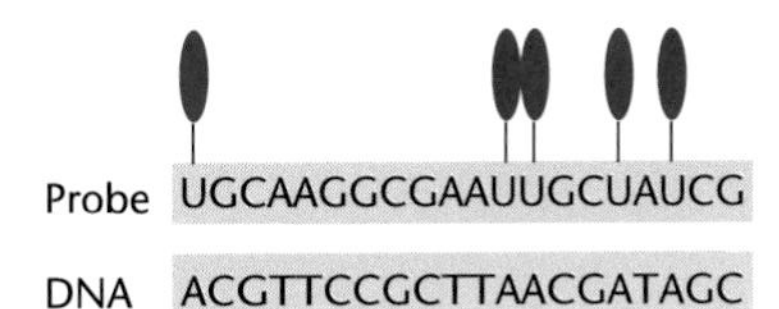

3 Antidigoxigenin antibody binds to digoxigenin molecules

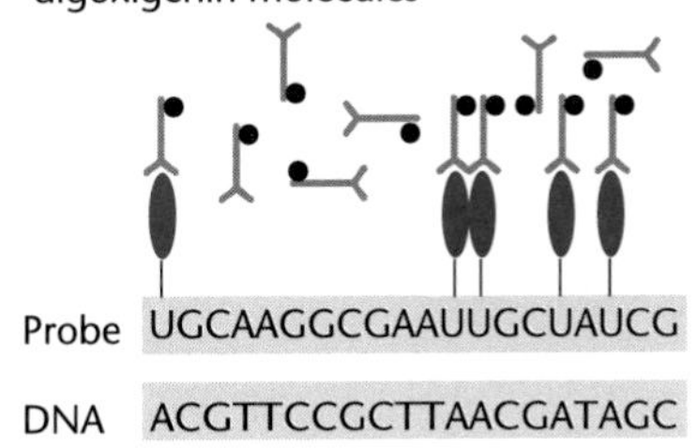

4 Unbound antibodies are washed from membrane

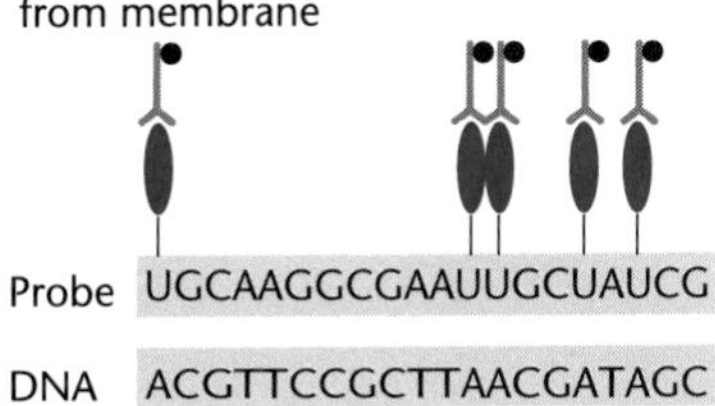

5 Alkaline phosphatase substrate is added, and precipitation reaction occurs

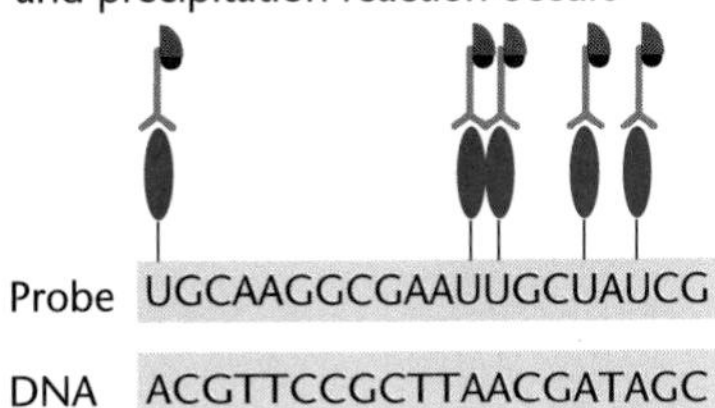

Color precipitate confirms that the probe has bound to the sequence of interest

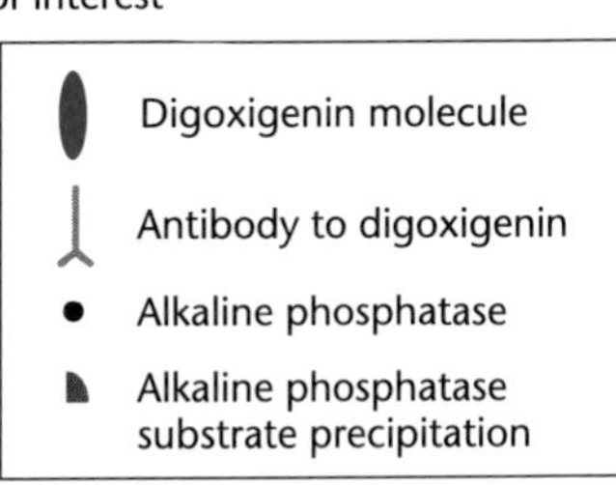

Figure 16.2 DNA probe hybridizes with its complement.

Lab Language

alkaline phosphatase
cross-link
denaturation
digoxigenin
DYZ1
D2S44
probe
RFLP
stringency

Laboratory Safety

Review the sections on physical, biological, and chemical hazards in "Laboratory Safety." Special considerations for this exercise:

hot agarose
gel electrophoresis
use of ethidium bromide
use of UV cross-linker
use of formamide
use of NBT and BCIP
10 M NaOH

Probe cannot find its complement

1 Probe is added to DNA

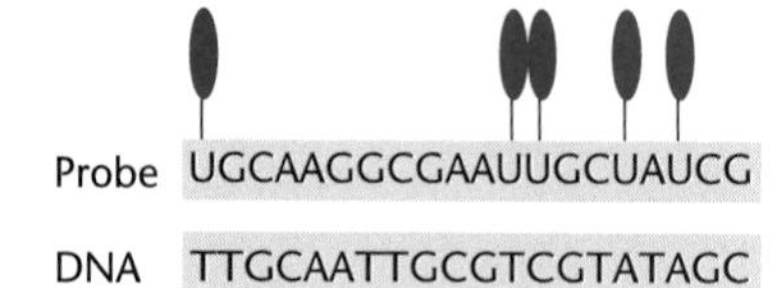

Probe UGCAAGGCGAAUUGCUAUCG

DNA TTGCAATTGCGTCGTATAGC

2 Unbound, excess probe is washed from membrane

DNA TTGCAATTGCGTCGTATAGC

3 Antidigoxigenin antibody is added

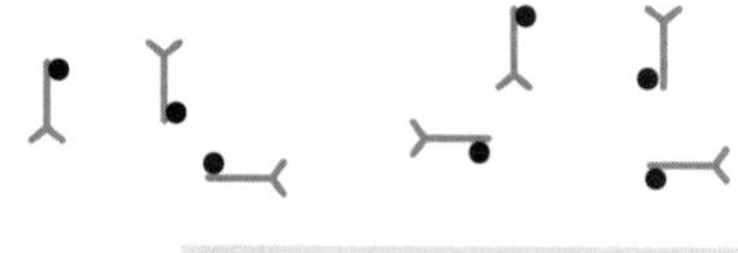

DNA TTGCAATTGCGTCGTATAGC

4 Unbound antibodies are washed from membrane

DNA TTGCAATTGCGTCGTATAGC

5 Alkaline phosphatase substrate is added, but no precipitation occurs

DNA TTGCAATTGCGTCGTATAGC

Absence of colored precipitate indicates that no complementary target DNA sequence is present

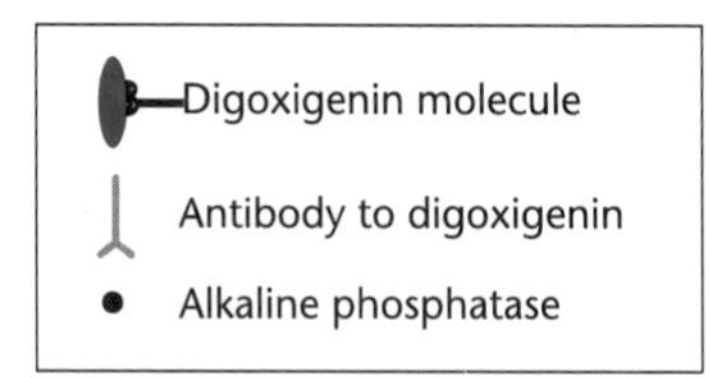

Figure 16.3 DNA probe does not hybridize to complementary DNA.

Laboratory Sequence

Day 1: Isolate genomic DNA and rehydrate overnight.

Day 2: Prepare and incubate genomic DNA restriction enzyme digest.
While incubating, prepare the agarose gel.

Day 3: Gel electrophoresis
While electrophoresing, prepare denaturation buffer and neutralization buffer.
Denature and neutralize gel.
While incubating, assemble the blotting materials.
Assemble Southern blot, and blot overnight.

Day 4: UV cross-link DNA to membrane.
Prehybridize membrane for 1 hour and hybridize with probe overnight.
While prehybridizing, prepare wash buffers.

Day 5: Wash excess probe from Southern blot membrane.
While washing, prepare developing reagents.
Develop Southern blot.

Materials

Genomic DNA Preparation

Class equipment and supplies

cell lysis solution (Tris-EDTA–sodium dodecyl sulfate [SDS]) (Gentra Systems, Inc.)
protein precipitation solution (ammonium acetate) (Gentra Systems, Inc.)
hydration solution (1× Tris-EDTA or from Gentra Systems, Inc.)
100% isopropanol
70% ethanol
clinical centrifuge
microcentrifuge
37°C water bath
65°C water bath
vortex

Materials per student

0.9% sterile NaCl solution (exercise 3)
plastic medicine cup
paper cup
15-ml conical centrifuge tube
microcentrifuge tubes
0.5- to 10-µl micropipette and sterile tips
10- to 100-µl micropipette and sterile tips

100- to 1,000-μl micropipette and sterile tips
1 aliquot of cell lysis solution
1 aliquot of ribonuclease (RNase)
1 aliquot of protein precipitation solution
1 aliquot of 100% isopropanol
1 aliquot of 70% ethanol
1 aliquot of hydration solution

Restriction Enzyme Digestion and Gel Electrophoresis

Class equipment and supplies
37°C water bath
microcentrifuge
ethidium bromide stain
0.8% agarose
10× Tris-borate-EDTA (TBE) buffer (or from exercise 3)
masking tape
dH_2O
*Hae*III restriction enzyme and buffer
Polaroid camera and film
UV transilluminator

Materials per pair of students
0.5- to 10-μl micropipette and sterile tips
1 aliquot of *Hae*III enzyme and 10× buffer
1 aliquot of loading dye
sterile microcentrifuge tubes
microcentrifuge tube rack
gel electrophoresis apparatus
power supply (two pairs of students will share this)
50-ml graduated cylinder
500-ml graduated cylinder
disposable exam gloves
permanent marking pen
beaker filled with crushed ice

Students will need to prepare
350-ml 1× TBE
50-ml 0.8% agarose gel

Southern Blotting

Class equipment and supplies
20× saline–sodium citrate (SSC)
5 M NaCl
1 M Tris, pH 7.0
10 M NaOH
Parafilm
UV cross-linker
Whatman 3MM filter paper
razor blade
blotting pads or paper towels
scissors

Materials per pair of students
10-ml pipette and dispenser bulb
100-ml graduated cylinder
250-ml graduated cylinder
500-ml graduated cylinder
blotting apparatus consisting of buffer container and 2 tiles
gloves
DNA marker

Students will need to prepare
150-ml denaturation buffer
300-ml neutralization buffer
300-ml 10× SSC
3 wicks approximately 10 by 30 cm
nylon membrane cut to gel size
3 pieces Whatman paper cut to gel size

DNA Probing and Detection

Class equipment and supplies
20× SSC
10% SDS
5 M NaCl
1 M Tris, pH 7.5
0.5 M $MgCl_2$
1 M Tris, pH 9.5
prehybridization buffer
formamide
sarcosine
heat block
42°C shaking water bath
50 to 65°C water bath
digoxigenin-labeled DNA probe
ice
balance
weigh paper

Materials per pair of students
0.5- to 10-μl micropipette and sterile tips
10- to 100-μl micropipette and sterile tips
100- to 1,000-μl micropipette and sterile tips
500-ml graduated cylinder
250-ml graduated cylinder
50-ml graduated cylinder
25-ml graduated cylinder
incubation and wash container
4 5-ml pipettes and dispenser bulb
1 2-ml pipette and dispenser bulb
2 50-ml test tubes

3 15-ml test tubes
gloves

Students will need to prepare
300 ml of 2× SSC + 0.1% SDS
50 ml of 0.5× SSC + 0.1% SDS
150 ml of wash buffer
15 ml of wash buffer + 2% blocking reagent
50 ml of equilibration buffer

Protocols

PROTOCOL 1 OBTAINING CHEEK CELLS

An isotonic solution, 0.9% NaCl, is used to maintain cell membrane integrity.

Cheek cells are constantly sloughed from the epithelium of the mouth.

1. Fill a plastic medicine cup with about 10 ml of sterile salt solution (0.9% NaCl). Rinse your mouth with the salt solution for at least 60 seconds. Expel the salt solution into a paper cup.
2. Carefully pour the salt solution, now containing your cheek cells, into a 15-ml tube. Close the tube and centrifuge it for 10 minutes. This will force your cells to the bottom of the tube. *Keep your cup and label your tube!*

The pellet is the solid part at the bottom of the tube.

The supernatant is the liquid portion obtained after centrifugation.

3. Pour the liquid back into the paper cup. Be careful not to disturb the cells at the bottom of the tube. Remove most of the supernatant, leaving 20 to 40 µl of liquid on the cell pellet.
4. Vortex the cells and supernatant to resuspend the cell pellet.

PROTOCOL 2 ISOLATION OF GENOMIC DNA

SDS, a detergent in the lysis solution, breaks open the cell membranes.

1. Add 600 µl of cell lysis solution and gently pipette up and down to lyse the cells. Transfer all of the liquid and cells to a clean microcentrifuge tube. Label your tube.
2. Incubate the tube at 37°C until the solution is homogeneous and no cell clumps remain visible. This step may be omitted if the solution resuspends without cell clumps.

RNase is an enzyme that destroys RNA.

3. Add 3 µl of RNase A solution to the cell lysate and incubate at 37°C for 30 minutes.
4. Cool the sample to room temperature.

Ammonium acetate causes the proteins to precipitate, leaving the DNA in solution.

5. Add 200 µl of protein precipitation solution and vortex vigorously for 20 seconds.
6. Microcentrifuge the sample at 14,000 rpm for 3 minutes to pellet the protein debris.
7. Pour the supernatant into a clean microcentrifuge tube, leaving the protein debris behind.

Isopropanol causes the DNA to precipitate out of solution.

Violent mixing will shear DNA into smaller pieces.

Orient the hinge on the microcentrifuge tube outward to assist in locating the tiny DNA pellet.

8. Add 600 µl of 100% isopropanol and mix by *gently* inverting the tube 50 times. White threads of DNA should form a visible clump.
9. Microcentrifuge the tube at 14,000 rpm for 1 minute. The DNA will be visible as a small white pellet.
10. Pour off the supernatant and drain the liquid on an absorbent towel.

11. Add 600 µl of 70% ethanol and invert the tube several times to wash the DNA.

12. Microcentrifuge the tube at 14,000 rpm for 1 minute.

13. Carefully pour off the ethanol. Pour slowly and carefully, watching the pellet so it does not pour out.

A 70% ethanol wash removes unwanted salts from the DNA.

Any residual ethanol will interfere with subsequent analysis.

14. Drain the liquid using an absorbent towel.

15. Allow the tube to air dry for at least 15 minutes.

16. Add 60 µl of DNA hydration solution and incubate at 65°C for 1 hour or overnight at room temperature to resuspend the DNA. Tap the tube occasionally to help disperse the DNA.

Do not pipette the DNA to mix, as this will cause shearing.

17. Use 16 µl of the isolated DNA in the restriction enzyme digestion for Southern blotting.

DNA quantity may be assessed by spectrophotometry or gel electrophoresis.

PROTOCOL 3 PREPARING THE RESTRICTION ENZYME DIGESTS

1. You will set up your digests in small tubes. In each tube, for each digest, you will need 16 µl of DNA, 2 µl of 10× restriction enzyme buffer, and 2 µl of 10-U/µl *Hae*III restriction enzyme.

It is anticipated that at least 8 to 10 µg of DNA will be digested.

2. Prepare duplicate samples.

3. Securely fasten the cap on each tube. To mix the reagents, pulse the tubes in the microcentrifuge.

*Hae*III *and* Hin*fI are used in forensics because their recognition sequences commonly flank variable number of tandem repeat (VNTR) regions.*

*Hae*III *recognizes the sequence GGCC/CCGG.*

4. Incubate the sample tubes in a 37°C water bath for at least 2 to 3 hours. The reaction may be allowed to incubate overnight.

5. Prepare the agarose gel while digests are incubating.

To ensure a complete digest, incubate genomic DNA overnight.

PROTOCOL 4 PREPARING 1× TBE AND THE AGAROSE GEL

A. Prepare 1× TBE Running Buffer

1. Prepare 350 ml of 1× TBE buffer for gel electrophoresis. Determine the volume of 10× TBE stock required; document calculations below.

TBE is usually made as a 10× stock solution.

TBE is used as the electrophoresis running buffer and to prepare the agarose gel.

2. Using a 50-ml graduated cylinder, measure the appropriate volume of 10× TBE and pour it into a 500-ml graduated cylinder.

3. Dilute the 10× TBE with distilled or deionized water to a final volume of 350 ml.

Do not use tap water: minerals will interfere with electrophoresis.

B. Preparing the Agarose Gel

1. Calculate the amount of agarose needed to prepare 50 ml of 0.8% concentration. Document calculations below.

2. Pour 50 ml of 1× TBE into a 125-ml Erlenmeyer flask. Add the powdered agarose.

While the agarose solution is cooling, begin protocol 5.

3. Microwave the agarose solution at the highest power for about 1 minute.

4. When the agarose solution is cool enough that the flask can easily be held in your hand, the agarose gel is ready to be poured.

PROTOCOL 5 CASTING THE AGAROSE GEL AND PREPARING THE GEL CHAMBER

A. Cast the Gel

Imperfections caused by moving the gel will skew your results.

1. Seal the ends of the casting tray and insert the comb in the notch closest to the end of the tray. Set the casting tray in a place on your lab bench where it will not be disturbed.

2. Pour the warm 0.8% agarose solution into the casting tray to a depth of about 5 mm. The gel will appear cloudy when it has solidified. Do not move the casting tray during solidification.

B. Prepare the Gel and Gel Chamber

The negative pole is usually color-coded black; the positive is red.

1. When the agarose gel has solidified, unseal the ends of the casting tray and place it in its correct position in the gel chamber. Make sure the comb is at the negative pole.

TBE is used to conduct the current through the gel.

2. Fill the chamber with just enough 1× TBE buffer to cover the entire surface of the gel.

3. *Gently* remove the comb, taking care not to tear the gel. Make sure the wells left by the comb are completely submerged. Add more buffer if necessary. The gel is now ready to be loaded with the DNA digests.

PROTOCOL 6 LOADING THE GEL AND SETTING UP THE GEL CHAMBER FOR ELECTROPHORESIS

Prepare duplicate gels, working with another laboratory group. One membrane will be probed with DYZ1, the other with D2S44.

Loading dye should constitute 10% of the volume of the digest.

1. Remove tubes from the water bath and add 2 µl of loading dye to each tube. Mix the loading dye with the contents of the tube by pulsing the tube in a centrifuge. Use a fresh tip for each tube.

2. Pipette each entire sample into a separate well in the gel. Use a fresh tip for each tube. Correlate and record sample identification with well number.

Markers are fragments of DNA of known sizes used to estimate the sizes of unknown DNA fragments.

3. Include a lambda-*Hin*dIII DNA marker (M) lane.

Gel 1

M 1 2 3 4

Gel 2

M 1 2 3 4

A stream of fine bubbles at the positive electrode is O_2 gas and at the negative electrode is H_2 gas, indicating current flow.

4. Slide the cover of the chamber into place and connect the electrical leads to the power supply so that the DNA migrates in the correct direction.

References and General Reading

Ausubel, F. M., R. Brent, R. E. Kingston, D. D. Moore, J. G. Seidman, J. A. Smith, and K. Struhl (ed.). 1994. *Current Protocols in Molecular Biology.* Wiley Interscience, New York, N.Y.

Boehringer Mannheim. DIG DNA Labeling and Detection Kit. Boehringer Mannheim Corp., Indianapolis, Ind.

Gentra Systems. Puregene DNA Isolation Kit. Gentra Systems, Inc., Minneapolis, Minn.

Kirby, L. T. 1992. *DNA Fingerprinting: an Introduction.* W. H. Freeman & Co., New York, N.Y.

Maniatis, T., E. F. Fritsch, and J. Sambrook. 1982. *Molecular Cloning: a Laboratory Manual.* Cold Spring Harbor Laboratory Press, Cold Spring Harbor, N.Y.

Nakahori, Y. 1986. A human Y-chromosome with specific repeated DNA family (DYZ1) consists of a tandem array of pentanucleotides. *Nucleic Acids Res.* **14:**7569–7573.

Nakamura, Y., M. Leppert, P. O'Connell, R. Wolff, T. Holm, M. Culver, C. Martin, E. Fujimoto, M. Hoff, E. Kumlin, and R. White. 1987. Variable number of tandem repeat (VNTR) markers for human gene mapping. *Science* **235:**1616–1622.

Odelberg, S. J., R. Plaetke, J. R. Eldridge, L. Ballard, P. O'Connell, Y. Nakamura, M. Leppert, J.-M. Lalouel, and R. White. 1989. Characterization of eight VNTR loci by agarose gel electrophoresis. *Genomics* **5:**915–924.

Southern, E. M. 1975. Detection of specific sequences among DNA fragments separated by gel electrophoresis. *J. Mol. Biol.* **98:**503–517.

Discussion

1. Why is the agarose gel placed in a denaturing solution before a Southern transfer?
2. What is the purpose of transferring the DNA out of the gel onto the membrane?
3. Why is the DNA cross-linked to the membrane?
4. Why are long, intact pieces of DNA required for a Southern blot? Describe your results if sheared pieces of DNA are used for a Southern blot.
5. Describe how the blot portion of Southern blotting works.

Observe the photo of the gel for the Southern blot.

6. What is your DNA marker lane? What is its purpose?
7. Observe your lane of digested DNA. How does this differ from undigested DNA?
8. Compare your digested DNA with the other lanes of digested DNA. Describe what you observe and why this is so.
9. What do you observe on your blotted membrane? Why do you think this is so?
10. How will the blotted portion of the experiment be analyzed?
11. In forensic testing intended to include or exclude a suspect, would you use more stringent or less stringent conditions? Why?
12. What effect will changing the salt concentration (SSC) have on stringency? Why?
13. What effect will changing the temperature have on stringency? Why?
14. Why is the probe labeled? Can you think of any other ways of doing this?
15. What DNA sequences make the best probes for use in forensic identification? Why?

Observe the probed and developed filter.

16. Locate your DNA marker lane. What do you observe? Why do you think this is so?

17. Locate the lane containing your digested DNA. What do you observe? Why do you think this is so?

18. Compare the lane of your digested DNA with the other lanes containing digested DNA. Do you observe any similarities? Do you observe any differences? Why do you think this is so?

19. Compare the results of the DYZ1 probe with the results of the D2S44 probe. Are there any similarities? Are there any differences? Why do you think this is so?

20. Enzyme A is used to cut DNA from individuals 1 to 5. It is electrophoresed in an agarose gel to separate the DNA by size. The DNA is then incubated with a probe, complementary to this region of DNA. Diagram the expected results. Can you identify different individuals? Why or why not?

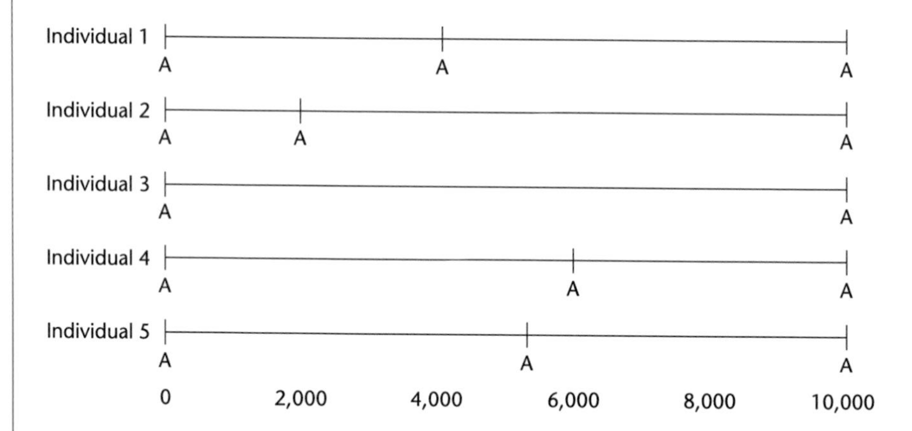

SECTION V

Protein Studies

EXERCISE 17

Quantitation of Protein

Key Concepts

1. Many areas of protein study require accurate, rapid methods to determine protein concentration.
2. Proteins may be quantitated by the Lowry assay, the Bradford assay, or the bicinchoninic acid (BCA) assay.
3. The Bradford assay is adaptable to standard as well as to micro assay protocols.

Goals

By the end of this exercise, you will be able to

- explain the use of colorimetry in protein assays
- discuss the relative benefits of each type of protein assay
- understand the reaction mechanism responsible for and perform the Bradford assay
- understand the necessity for and prepare a standard curve of known bovine serum albumin (BSA) concentrations
- determine the protein concentration in unknown samples

Introduction

Proteins are the expressions of genes. They provide for the structural framework of the cell, regulation of cellular function, signal transduction, and enzyme activity. Protein analysis, necessary for an understanding of protein function, often begins with quantitative measurement of total protein concentration in a biological sample. Probably the most accurate method for determining protein concentration would be to perform acid hydrolysis of the protein, then analyze the amino acid composition. This procedure would be lengthy and painstaking to perform. For initial protein studies, simpler, yet easily replicable methods to assess protein concentration are highly desirable.

Three different methods are frequently employed in many laboratories. These are the Lowry method, the Bradford method, and the BCA method. All three of these methods are sensitive to the amino acid composition of the protein in question and make it difficult to obtain absolute concentration. However, these methods are so widely used and relatively simple to perform that they have become well accepted as means to determine protein concentration, especially in protein mixtures or crude extracts.

In each of the aforementioned methods, protein concentration is determined by comparison of unknown samples with a standard curve generated by performing the desired protocol using known concentrations of a protein standard solution. BSA, fraction V, and bovine gamma globulin (BGG) are the most widely used protein assay standards.

The Lowry Method

Developed in 1951, the Lowry protocol is the oldest method for the assay of protein concentration. It is based on the biuret reaction and the poorly understood Folin-Ciocalteu reaction. Cu^{2+} is reduced to Cu^{+} by reacting with peptide bonds under alkaline conditions. In turn, Cu^{+} reacts with the Folin reagent, catalyzing oxidation of aromatic amino acids with concomitant reduction of phosphomolybdotungstate to heteropolymolybdenum blue. The intensity of blue color development can be measured by absorbance at either 750 nm (low expected protein concentration, below 500 μg/ml) or 550 nm (high protein concentration, 100 to 2,000 μg/ml).

Color development is partly dependent on tyrosine and tryptophan content of the protein. Despite this, sensitivity is reasonably constant from protein to protein. The lower limit of sensitivity is about 0.01 mg of protein per ml. The assay works best when protein concentrations are between 0.01 and 1.0 mg of protein per ml. The Lowry method is more accurate than the Bradford assay, but the presence of reducing agents, including reducing sugars, does interfere with Lowry analysis.

The Bradford Method

The Bradford protocol is based on the binding of Coomassie blue G250 dye to protein. The dye binds to protein tertiary structure and to specific amino acids, especially to arginyl and lysyl residues of proteins, although Coomassie blue will not bind to the free amino acids, arginine and lysine. Free Coomassie blue can exist in several ionic forms, depending on the pH of the reagent solution. Cationic red and green forms of the dye predominate in the acidic reagent solution, but it is the more anionic blue form that binds to protein. Therefore, protein concentration can be determined by measuring the absorbance of the solution at 595 nm. Although binding of Coomassie blue to protein shifts the absorbance maximum of the blue dye form from 590 to 620 nm, the latter wavelength is not used because the green form of the dye, with a maximum absorbance at 650 nm, would interfere. Measurement is standardized at 595 nm, the wavelength formerly believed to be the absorbance maximum of the blue form.

The Bradford method is faster than either the BCA or the Lowry method and is more sensitive than the Lowry method. Protein concentrations in the ranges of 100 μg/ml to 1.5 mg/ml (standard protocol) or 1 to 25 μg/ml (micro protocol) can be detected. Proteins with as low a molecular weight as 3,000 can be measured. The Bradford method has the advantage of being compatible with reducing agents. Additionally, less interference from nonprotein components of biological samples occurs with the Bradford method. One disadvantage, however, is that detergents do interfere with the Bradford assay. Many workers overcome this by incorporating the detergent that is present in the protein preparations into a blank control and into the calibration standards. Another disadvantage of the Bradford method is that it exhibits more protein-to-protein variability than either the Lowry or BCA method.

The BCA Method

The BCA method is similar to the Lowry method in that Cu^{2+} is reduced to Cu^{+} by reacting with peptide bonds under alkaline conditions, but the presence of Cu^{+} is detected by reaction with BCA. The resulting color product, intensely purple, is dependent on protein concentration and incubation time. Absorbance is read at 562 nm. One advantage of the BCA assay is that only one step is required, compared with the two steps necessary for the Lowry method. In addition, the BCA assay is as accurate as the Lowry assay and exhibits less protein-to-protein variability than the Bradford assay. However, reducing agents, including reducing sugars, do interfere with the BCA assay.

The BCA assay is more sensitive than the Lowry assay. Depending on whether a standard protocol or a micro protocol is performed, BCA assays can detect proteins in the range of 5 µg/ml to 2 mg/ml or 0.5 to 20 µg/ml. Proteins with as low a molecular weight as 2,000 can be measured.

This laboratory exercise will explore the use of the Bradford method (standard test tube protocol), since it is reliable and the quickest, easiest method to estimate protein concentration in crude biological samples.

Lab Language

BCA assay
Bradford assay
colorimetry
Lowry assay
standard curve

Laboratory Safety

Review the sections on physical, biological, and chemical hazards in "Laboratory Safety." Special considerations for this exercise:

use of spectrophotometer

Materials

Bradford Method for Protein Quantitation

Class equipment and supplies

spectrophotometer
clinical centrifuge
microcentrifuge
vortex mixer
BSA fraction V standard concentrate, 2 mg/ml in 0.9% NaCl
Bradford reagent
dH_2O
1% egg albumin solution
1% chicken liver homogenate
1% egg yolk homogenate
1% egg white homogenate
1% fish egg (caviar) homogenate
1% tofu (firm) homogenate

Materials per pair of students

linear graph paper
250-ml beaker
disposable cuvettes
test tubes
1-ml and 5-ml pipettes
pipette aids for 1- and 5-ml pipettes
disposable exam gloves
0.5- to 10-µl micropipette and sterile tips
100- to 1,000-µl micropipette and sterile tips

Students will need to prepare

BSA protein assay standards
suitable dilutions of crude protein samples

Protocols

PROTOCOL 1 PREPARATION OF THE DILUTED BSA STANDARDS

1. Obtain eight microcentrifuge tubes. *Wear gloves.* Label the tubes with the following final BSA concentrations and letters:

   2,000 µg/ml, S (stock)
   1,500 µg/ml, A
   1,000 µg/ml, B
   750 µg/ml, C
   500 µg/ml, D
   250 µg/ml, E
   125 µg/ml, F
   25 µg/ml, G

Standards need to be prepared fresh just before use.

2. Use the following chart to serially dilute the BSA stock (S) solution, 2,000 µg/ml, to achieve the desired final concentrations.

Standards must be diluted in the same diluent as the samples to be assayed.

Volume of BSA	Volume of dH_2O	Final BSA concentration
300 μl S	None	2,000 μg/ml
375 μl S	125 μl	1,500 μg/ml = A
325 μl S	325 μl	1,000 μg/ml = B
175 μl A	175 μl	750 μg/ml = C
325 μl B	325 μl	500 μg/ml = D
325 μl D	325 μl	250 μg/ml = E
325 μl E	325 μl	125 μg/ml = F
100 μl F	400 μl	25 μg/ml = G

3. With the 100- to 1,000-μl pipette and a clean pipette tip in place, pipette the required quantity of dH_2O into each tube. Obtain a fresh tip to pipette the stock BSA solution. Vortex or finger-flick each tube to mix well.

4. Use a clean tip for each of the subsequent serial dilutions, mixing each before proceeding to the next.

PROTOCOL 2 DILUTION OF UNKNOWN SAMPLES

Hint: convert the concentration of the unknown sample from percentage to micrograms per milliliter.

1. Estimate the possible target concentration of the unknown sample, and dilute the sample with dH_2O to provide "guesstimates" of between 100 and 1,000 μg of protein in 1 ml. Document calculation below. For example, egg white is approximately 10.54% protein, and egg yolk is about 17.4% protein. How would you proceed?

2. You may wish to prepare and test several dilutions, such as 1/10, 1/100, 1/1,000, as well as undiluted samples.

3. Measure the appropriate quantity of the unknown sample and dH_2O into a clean tube for each dilution.

PROTOCOL 3 MEASUREMENT OF PROTEIN CONCENTRATION

The absorbance at 595 nm is dependent on the temperature of the reagent.

Cold reagent will result in lower readings.

1. The Bradford reagent must be at room temperature. Mix the solution just before use by gently inverting the bottle several times. *Do not shake!*

2. Pipette 50 μl of each standard and of each unknown sample into appropriately labeled test tubes or disposable cuvette tubes.

3. Pipette 50 µl of dH_2O into a labeled test tube or cuvette tube to prepare a reagent blank.

4. Add 1.5 ml of Bradford reagent to each tube. Mix well, and incubate at room temperature for 2 minutes.

5. Zero the spectrophotometer, at 595 nm, with a distilled water reference.

Review exercise 11 for use of the spectrophotometer.

6. Measure the absorbance at 595 nm of each tube versus that of the water reference, beginning with the reagent blank. Record your data in Table 17.1.

Measure the absorbance of the reagent blank between each reading of standard curve and unknown sample.

7. Subtract the average 595 nm reading for the reagent blank from the 595 nm reading for each standard and each unknown sample. This will provide a corrected reading.

The average reading for the reagent blank is used to compensate for changes in temperature during the procedure.

8. Prepare a standard curve by plotting the corrected 595 nm reading (y axis) for each BSA standard versus its concentration in micrograms per milliliter (x axis).

9. Determine the protein concentration for each unknown sample by using its corrected 595 nm reading to extrapolate from the standard curve.

Table 17.1 Bradford assay data: standards and unknown samples

Name/concentration of sample	Reading of reagent blank at 595 nm	Reading of sample at 595 nm	Corrected reading of sample[a]
	Average reagent blank reading =		

[a]Sample reading minus average reagent blank reading.

References and General Reading

Bradford, M. M. 1976. A rapid and sensitive method for the quantitation of microgram quantities of protein utilizing the principle of protein-dye binding. *Anal. Biochem.* **72:**248–254.

Lowry, O. H., N. J. Rosebrough, A. L. Farr, and R. J. Randall. 1951. Protein measurement with the Folin phenol reagent. *J. Biol. Chem.* **193:**265–275.

Smith, P. K., R. I. Krohn, G. T. Hermanson, A. K. Mallia, F. H. Gartner, M. D. Provenzano,

E. K. Fujimoto, N. M. Goeke, B. J. Olson, and D. C. Klenk. 1985. Measurement of protein using bicinchoninic acid. *Anal. Biochem.* **150:**76–85.

Walker, J. M. (ed.). 1996. *The Protein Protocols Handbook.* Humana Press, Totowa, N.J.

Discussion

1. Suggest some reasons why scientists may need to perform protein quantification studies.
2. What are the advantages of the Lowry method? Disadvantages?
3. What are the advantages of the BCA method? Disadvantages?
4. What are the advantages of the Bradford method? Disadvantages?
5. Which of the unknown samples appeared to possess the highest concentration of protein? The lowest?
6. Are your results as you expected? Why? Why not?

EXERCISE 18

Sodium Dodecyl Sulfate-Polyacrylamide Gel Electrophoresis and Coomassie Blue Staining of Protein

Key Concepts

1. Sodium dodecyl sulfate-polyacrylamide gel electrophoresis (SDS-PAGE) denatures proteins and separates them based on molecular weight.
2. Proteins may be positively or negatively charged. SDS is added to give them a net negative charge for electrophoresis.
3. Coomassie blue staining easily stains all proteins in an acrylamide gel for visualization.

Goals

By the end of this exercise, you will be able to

- discuss the purpose of each of the components of Laemmli sample buffer
- describe and perform SDS-PAGE
- understand Coomassie blue staining of proteins and the results obtained

Introduction

Protein Structure

Proteins are the expressed products of genes. Their primary structure is a result of translation of triplet nucleotide codons through a messenger RNA intermediate into an amino acid sequence. This primary sequence, however, does not tell the complete story of the protein's ultimate structure and function. The secondary structure of proteins occurs when chains of amino acids arrange themselves into alpha-helices or beta-pleated sheets. Combinations of secondary structures, usually containing fewer than 150 amino acids, organize into protein domains constituting the tertiary structure of the protein. Finally, the multiple domains of the protein are strung together by polypeptides to form the globular protein. Covalent intrachain bonds, such as disulfide bonds between cysteine residues, may form between polypeptides to stabilize three-dimensional protein structures. To be functional, some proteins must form aggregates of homoproteins (proteins with identical subunits) or heteroproteins (proteins with different subunits).

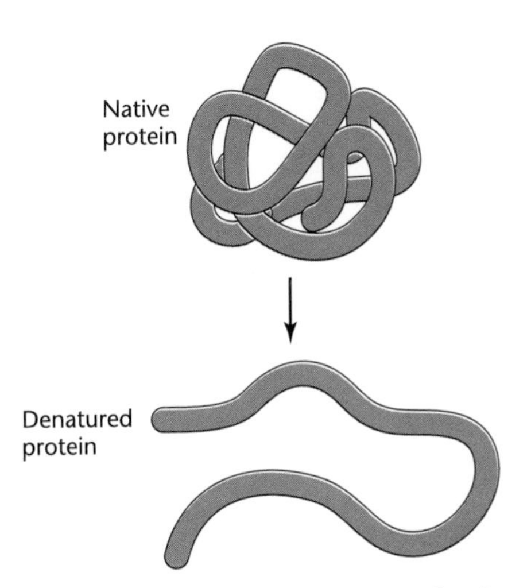

Figure 18.1 The globular structure of native proteins is dissociated by LSB components and heat treatment.

SDS-PAGE

One-dimensional SDS-PAGE is most often used to denature proteins and separate them based on size. Preparation of material for SDS-PAGE analysis is simple. Cells are lysed in Laemmli sample buffer (LSB), then heated to boiling. This denatures the proteins (Fig. 18.1), breaks down the subunits, and dissociates protein-protein interactions. Since proteins may be positively or negatively charged,

SDS, an anionic detergent, is included in the LSB. SDS coats all the proteins, giving them a net negative charge. The proteins will travel in the polyacrylamide gel toward the positive electrode during electrophoresis. A second key component of the sample buffer is dithiothreitol (DTT). DTT reduces disulfide bonds to assist with denaturation of the proteins into their subunits. LSB also contains bromophenol blue and glycerol. Glycerol makes the protein sample dense, to simplify loading the samples into the wells of the gel. Bromophenol blue serves to assist in monitoring the progress of the electrophoresis. SDS is included as a component in the electrophoresis running buffer to maintain negative charges on the proteins. Proteins prepared in this way are ready for electrophoresis.

Like agarose gel electrophoresis, altering the concentration of acrylamide changes the pore size of the gel. Larger pore sizes, obtained with lower acrylamide concentration, resolve larger proteins best. Smaller pore sizes, achieved with higher acrylamide concentration, resolve smaller proteins best. The table below serves as an approximate guide. The average molecular mass of one amino acid is 110 Da (daltons). A 10-kDa (kilodalton) protein is approximately 91 amino acids in length.

Acrylamide concentration	5%	7%	10%	15%
Range of proteins resolved	57–212 kDa	36–94 kDa	16–68 kDa	12–43 kDa

Two-dimensional gel electrophoresis is a very powerful method for separating complex mixtures of as many as 1,500 different proteins in a single gel. A one-dimensional gel resolves only about 100 proteins in a similar sample. Proteins in a two-dimensional gel are first separated by their isoelectric point in a thin tube gel. This first-dimension gel is then placed on top of an SDS-polyacrylamide slab gel and electrophoresed again. Proteins are resolved based on molecular weight in the second dimension. Since charge and molecular weight are both monitored, two-dimensional gels are useful in assessing small changes in proteins.

Analysis of proteins in their native, nondenatured state is important when assessing the function of a specific protein. One-dimensional gel electrophoresis in the absence of SDS and DTT keeps protein subunits intact and maintains protein conformation, thereby preserving biological activity of the protein. The true molecular weight of the protein may not be determined because the separation is based on molecular size, shape, and charge. Nondenaturing gels can be used to determine the purity of a protein complex or multimeric protein, which should migrate as a single protein band.

Once proteins are resolved by SDS-PAGE or another electrophoretic method, they must be detected and analyzed. This may be carried out by several methods. Stains, such as Coomassie blue or silver stain, are used to detect any and all proteins directly in the SDS-polyacrylamide gel. Alternatively, the proteins may be blotted to nitrocellulose, and a protein of interest can be identified and detected with a specific antibody.

Protein Staining

One method of visualizing proteins directly in an acrylamide gel is to immerse the gel in a solution of stain. Two stains are commonly used, Coomassie blue or silver. They differ in ease of use, speed of result, and sensitivity. Coomassie blue staining is easier and faster than silver staining. Coomassie blue may be prepared and stored for long periods of time before use. The reagents for silver staining must be prepared more frequently. Silver staining is more expensive than Coomassie blue staining. Silver staining, however, is sensitive to protein concentrations in the 2- to 5-ng range. Coomassie blue staining requires protein concentrations of 0.3 to 1 μg.

Lab Language

denature
disulfide bond
heteroprotein
homoprotein
isoelectric
native
one-dimensional gel electrophoresis
SDS-PAGE
two-dimensional gel electrophoresis

Laboratory Safety

Review the sections on physical, biological, and chemical hazards in "Laboratory Safety." Special considerations for this exercise:

working with microbes
boiling samples
gel electrophoresis
SDS-PAGE: acrylamide, *N,N,N',N'*-tetramethylethylenediamine (TEMED), DTT

Materials

Class equipment and supplies
culture of *Escherichia coli*
culture of *Saccharomyces cerevisiae*
plant tissue
Euglena culture
30% acrylamide-bisacrylamide, 37.5:1
1.5 M Tris, pH 8.8
10% SDS
10% ammonium persulfate (APS)
TEMED
1.0 M Tris, pH 6.8
Pasteur pipettes and bulbs
clinical centrifuge
microcentrifuge
heat block or boiling water bath
2× LSB
protein marker in 2× LSB
10× SDS running buffer
Coomassie blue stain
destain
gel-drying apparatus and drying film
dH_2O
Kimwipes

Materials per pair of students
plastic medicine cup
4 disposable 15-ml tubes
vertical gel electrophoresis system (glass plates, spacers, comb, and chamber)
gloves
sterile 0.9% NaCl solution (exercise 3)
paper cup
microcentrifuge tubes
1 aliquot (600 µl) of 2× LSB
0.5- to 10-µl micropipette and sterile tips
100- to 1,000-µl micropipette and sterile tips
power supply capable of 250 volts (shared by 2 to 4 groups)
500-ml graduated cylinder
50-ml graduated cylinder
squirt bottle containing dH_2O

Students will need to prepare
10% polyacrylamide gel
5% stacking gel
300-ml 1× SDS running buffer

Protocols

PROTOCOL 1 PREPARATION OF CHEEK CELL LYSATE

1. Fill a plastic medicine cup with about 10 ml of sterile salt solution (0.9% NaCl). Rinse your mouth with the salt solution for about 10 seconds. Expel the salt solution into a paper cup.

2. Carefully pour the salt solution, now containing your cheek cells, into a 15-ml tube labeled with your name. Close the tube and centrifuge for 10 minutes. This will force your cells to the bottom of the tube. *Save your cup!*

The supernatant is the liquid portion obtained after centrifugation.

3. Pour the supernatant back into the paper cup. Be careful not to disturb the cells at the bottom of the tube. Remove as much liquid as possible from the cell pellet.

The pellet is the solid part at the bottom of the tube.

LSB contains SDS, which gives all proteins a net negative charge, and DTT, which breaks disulfide bonds.

LSB also contains glycerol to make the sample heavy for loading purposes and bromophenol blue to monitor the progress of electrophoresis.

4. Add 100 µl of 2× LSB to the cell pellet and gently pipette up and down to mix well. Transfer all the contents to a 1.5-ml microcentrifuge tube. Label your tube.

Heating assists with protein denaturation.

5. Heat all samples at 100°C for 5 minutes, *just before loading the gel.*

PROTOCOL 2 PREPARATION OF PLANT CELL LYSATE

Tissue must be dissociated into single cells.

1. Collect a fresh, tender leaf. Cut a 1-cm-square piece of tissue from it.
2. Place the leaf tissue in a microcentrifuge tube. Crush it into a paste with a toothpick or a small pestle.
3. Add 100 µl of 2× LSB and vortex to mix.
4. Incubate the tube at room temperature for 5 minutes, flicking occasionally to mix.

Large pieces of plant debris will interfere with electrophoresis.

5. Microcentrifuge at 10,000 rpm for 3 minutes to separate particulate material from leaf protein extract.
6. Using a pipette, transfer the supernatant into a clean, labeled microcentrifuge tube.
7. Heat all samples at 100°C for 5 minutes, *just before loading the gel.*

PROTOCOL 3 PREPARATION OF BACTERIAL CELL LYSATE

Work carefully with microbes.

1. Obtain an overnight broth culture of *E. coli.*
2. Label a microcentrifuge tube. Transfer approximately 1.5 ml of the culture into the tube, using a Pasteur pipette.
3. Pulse the tube in the microcentrifuge for 10 seconds to pellet the bacteria.
4. Carefully remove most of the supernatant by pouring it into a container of disinfectant.
5. Add 100 µl of 2× LSB and gently pipette up and down to mix.
6. Heat the tube and its contents at 100°C for 5 minutes, *just before loading the gel.*

PROTOCOL 4 PREPARATION OF YEAST CELL LYSATE

1. Obtain an overnight broth culture of *S. cerevisiae.*
2. Label a microcentrifuge tube. Transfer approximately 1.5 ml of the culture into the tube, using a Pasteur pipette.
3. Pulse the tube in the microcentrifuge for 10 seconds to pellet the yeast.

4. Carefully remove most of the supernatant by pouring it into a container of disinfectant.

5. Add 100 µl of 2× LSB and gently pipette up and down to mix.

6. Heat the tube and its contents at 100°C for 5 minutes, *just before loading the gel.*

PROTOCOL 5 PREPARATION OF *EUGLENA* LYSATE

1. Obtain a fresh, pure liquid *Euglena* culture.

2. Label a microcentrifuge tube. Transfer approximately 1.5 ml of the culture into the tube, using a Pasteur pipette.

3. Pulse the tube in the microcentrifuge for 10 seconds to pellet the euglena.

4. Carefully remove most of the supernatant by pouring it into a container of disinfectant.

5. Add 50 µl of 2× LSB and gently pipette up and down to mix.

6. Heat the tube and its contents at 100°C for 5 minutes, *just before loading the gel.*

PROTOCOL 6 PREPARATION OF THE SDS-POLYACRYLAMIDE GEL

1. **Wearing gloves,** clean the glass plates with soap and water. Rinse well.

Clean plates help prevent bubble formation when acrylamide gels are poured.

2. Assemble the gel apparatus as shown in Fig. 18.2. The short glass plate should face outward, and the longer plate should face inward. The clamp knobs should face inward.

Align the glass plates and spacers on a flat surface.

The glass plates and spacers should form a watertight seal against the gasket.

3. Pipette water between the glass plates. Water must not leak from the assembly. Pour the water out, and blot with a paper towel.

Water is used to check the assembly to avoid wasting or contaminating the work area with acrylamide.

4. **Wearing gloves,** mix the reagents together for 5 ml of a 10 or 15% polyacrylamide resolving gel in a disposable tube. Add reagents in the order listed in the table below.

Acrylamide is a neurotoxin: ***avoid skin contact.***

APS initiates polymerization.

TEMED speeds up the polymerization process.

Reagent	10% Gel	15% Gel
dH_2O	2.00 ml	1.15 ml
1.5 M Tris, pH 8.8	1.25 ml	1.25 ml
30% acrylamide-bisacrylamide	1.70 ml	2.50 ml
10% SDS	50 µl	50 µl
10% APS	50 µl	50 µl
TEMED	5 µl	5 µl

5. Immediately "pour" the gel. Hold the comb outside the glass plates. Using a Pasteur pipette, fill the plates to about 5 mm below the lowest level of the comb.

6. Make sure there are no bubbles in the gel. Sharply tap the glass to force any bubbles to the top of the gel.

7. Add a thin layer of isopropanol or isobutanol to the top of the gel.

Polymerization is inhibited by exposure to air.

8. Save your tube containing the acrylamide solution. When the solution in the tube has solidified, the gel has also solidified, and the stacking gel can be poured.

Begin preparation of your samples while the gel is solidifying.

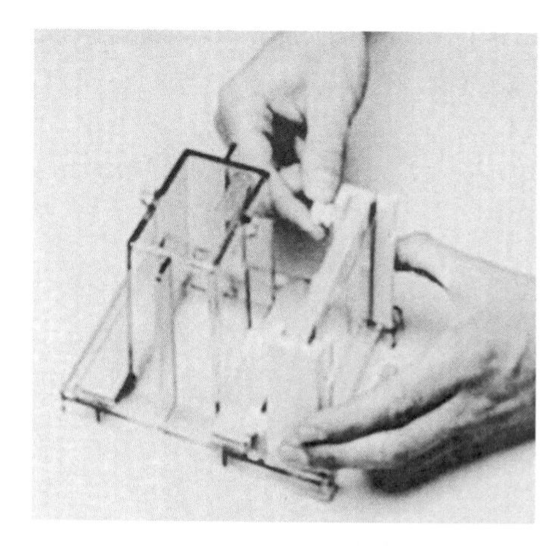

Figure 18.2 Assembly of gel apparatus. (Reprinted from Bio-Rad Laboratories, Hercules, Calif., with permission.)

9. Decant the alcohol from the 10 or 15% polyacrylamide gel and rinse several times with dH_2O to remove all the alcohol. Blot with a paper towel.

The stacking gel ensures that all proteins enter the resolving gel together.

10. **Wearing gloves,** prepare 2.5 ml of the 5% stacking gel in a disposable tube. Add reagents in the order listed in the table below.

Reagent	Volume
dH_2O	1.70 ml
1.0 M Tris, pH 6.8	313 µl
30% acrylamide-bisacrylamide	416 µl
10% SDS	25 µl
10% APS	13 µl
TEMED	2.5 µl

11. Pour the stacking gel between the glass plates using a Pasteur pipette. Insert the comb between the glass plates, taking care not to get bubbles stuck underneath the teeth.

12. Save the extra stacking gel solution to determine when the stacking gel has solidified.

PROTOCOL 7 SDS-PAGE

1. Transfer your gel to the electrophoresis chamber. Two groups will share an electrophoresis chamber to electrophorese two gels simultaneously.

2. Prepare 300 ml of 1× SDS running buffer from the 10× stock solution provided. One gel chamber requires 300 ml.

3. Add enough 1× SDS running buffer to the inside chamber so that it covers the wells.

4. Pour enough running buffer into the bottom of the chamber to make contact with the bottom of the gel.

The protein marker contains proteins of known sizes for comparison with unknown samples.

5. Obtain an aliquot of protein marker.

6. Heat all samples, including the marker, at 100°C for 5 minutes *immediately before loading.*

7. Load 10 μl of sample into each well as follows: lane 1, protein marker; lane 2, cheek cell lysate; lane 3, yeast cell lysate; lane 4, bacterial cell lysate; lane 5, *Euglena* lysate; lane 6, plant lysate.

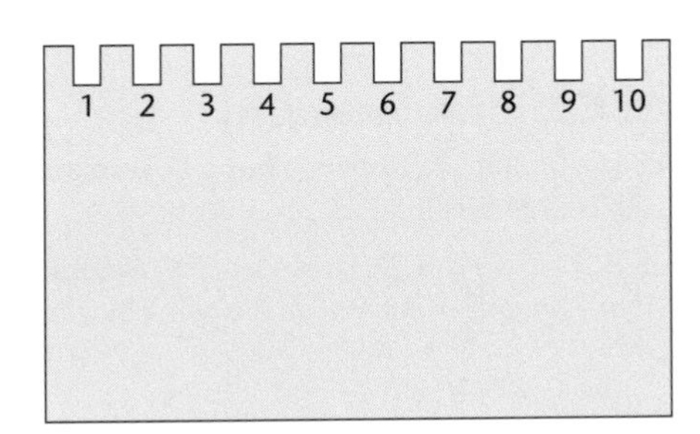

8. Place the cover on the gel apparatus and plug the electrodes into the power supply.

9. Turn on the power supply and electrophorese at 200 volts for about 1 hour, until the bromophenol blue dye front reaches the bottom of the gel.

A fine stream of bubbles from the electrodes indicates current flow.

PROTOCOL 8 COOMASSIE BLUE STAINING AND DESTAINING

When the electrophoresis is complete, **turn off the power supply. Wearing gloves,** carefully take the apparatus apart.

1. Remove the gel apparatus from the buffer chamber, and remove the plates containing the gel from the holder.

The glass plates are very fragile.

2. Remove the spacers and use one of the spacers to gently pry the glass plates apart. Take care not to rip the gel.

3. Use a squirt bottle of dH_2O to keep the gel moist and to prevent tears.

4. Cut off a small corner of the gel at the lower edge of the marker lane to provide for orientation.

5. Stain the gel in Coomassie blue by incubating at room temperature for 30 minutes.

Coomassie blue stains all proteins present in the acrylamide gel.

6. Destain the gel by incubating in destain solution 30 minutes to overnight at room temperature. Crumple a Kimwipe and place it in the corner of the container, in the destain solution, to absorb the stain.

The Kimwipe scavenges stain out of the solution and binds it.

PROTOCOL 9 DRYING THE POLYACRYLAMIDE GEL

Once the Coomassie blue-stained gel has been sufficiently destained, it may be air dried between sheets of gel-drying film as a permanent record of the proteins studied.

1. Soak two sheets of gel-drying film in water.

2. Place a sheet of film on top of one of the gel frames.

3. Gently place the acrylamide gel on the film. Use a squirt bottle with water to remove bubbles from between the gel and the film and to keep the gel moist.

Keep the gel and your gloves wet; the gel may stick to your gloves, causing tears.

4. Lay the second sheet on top of the acrylamide gel.

5. Carefully sandwich the gel between the bottom and top pieces of the film. Gently push bubbles to the sides with your fingers. *Make sure all of the air bubbles are removed from between the two layers of film.*

6. Align the second gel frame on top of the first and securely clamp the film-gel-film sandwich between the two frames.

7. Air dry the gel overnight.

References and General Reading

Alberts, B., D. Bray, J. Lewis, M. Raff, K. Roberts, and J. D. Watson. 1989. *Molecular Biology of the Cell*, 2nd ed. Garland Publishing, Inc., New York, N.Y.

Ausubel, F. M., R. Brent, R. E. Kingston, D. D. Moore, J. G. Seidman, J. A. Smith, and K. Struhl (ed.). 1994. *Current Protocols in Molecular Biology.* Wiley Interscience, New York, N.Y.

Discussion

1. LSB contains reagents that assist with protein denaturation. List each reagent and its function.
2. What information does an SDS-polyacrylamide gel provide and why is this useful?
3. What other types of gel analysis might be performed on a protein and why?
4. What is Coomassie blue? How does it function?
5. Examine your Coomassie blue-stained gel. Do the lysates contain any common proteins? Justify your answer.
6. Do the lysates contain different proteins? Justify your answer.

EXERCISE 19

Western Blotting and Immunodetection

Key Concepts

1. Sodium dodecyl sulfate-polyacrylamide gel electrophoresis (SDS-PAGE) may be used to analyze cell- and tissue-specific protein expression.
2. For identification of a particular protein resolved by SDS-PAGE, the proteins are transferred out of the gel and immobilized onto a nitrocellulose membrane in a process called Western blotting.
3. Antibodies are used to detect specific proteins in a Western blot. These antibodies are generated by immunizing an animal with the protein of interest.

Goals

By the end of this exercise, you will be able to

- describe the process of Western blotting
- understand the use of primary and secondary antibodies in immunodetection
- discuss the applications of Western blotting
- understand the purpose of tags on antibodies and discuss how alkaline phosphatase acts as a tag

Introduction

Proteins are the products of genes and provide cellular structure and function. Analysis of proteins is therefore central to understanding biological processes. Cells, even simple one-celled organisms, are composed of complex mixtures of many proteins, some with similar molecular weights. Assessments of protein expression, quantity, and tissue distribution are only some of the analyses required to more fully understand a specific protein's function. Data derived from gel electrophoresis, Western blotting, and immunodetection of proteins provide one mechanism for protein analysis. This laboratory exercise will couple SDS-PAGE analysis with Coomassie blue staining (see exercise 18), Western blotting, and immunodetection for protein analysis.

Western Blotting

Western blotting involves transfer and immobilization of proteins from a completed polyacrylamide gel to nitrocellulose, a paperlike membrane. It is based on the same principles as Southern blotting (see exercise 16). Buffer is drawn through the polyacrylamide gel to the nitrocellulose membrane by capillary action. The proteins are transferred to the membrane in a pattern mirroring their resolution in the gel. After E. M. Southern developed the blotting technique for DNA, compass points were chosen as names for other blotting techniques: Northern blotting for RNA and Western blotting for proteins.

Immunodetection

Western blot analysis is coupled with immunodetection, that is, using an antibody to specifically bind to the protein of interest (Fig. 19.1). Antibodies are protein molecules produced by the B cells of the immune system. Each is made in response to a specific foreign protein, called an antigen. An antibody may be generated against any antigen, such as bacterial, viral, yeast, or parasite components. The key feature of this response is that the antigen must be recognized as a foreign protein, not as a self protein. In the infrequent cases in which antibodies against self are made, conditions such as arthritis, lupus erythematosus, multiple sclerosis, or other autoimmune diseases result. Antibody response is very specific, because each antibody is generated against a single portion of the foreign protein, called an antigenic determinant or epitope. The human body can potentially synthesize several hundred different antibodies against a foreign protein, each specific for a different part of the protein structure.

Scientists have learned how to manipulate the specific antibody response. By injecting mice, rats, or sometimes rabbits with any protein of interest, including human proteins, animals can be induced to produce antibodies useful in research and diagnostics. The laboratory animal produces large amounts of a specific antibody, under carefully controlled conditions. When the serum, the liquid component of the blood, is collected from these animals, it is termed a polyclonal antiserum. Polyclonal means that the serum contains many antibodies of different specificities, each produced by a separate B-cell clone. These may include antibodies to different epitopes or regions of the protein of interest and will also include antibodies to any foreign agent that the animal has encountered in its lifetime.

A monoclonal antibody is a much more specific reagent. It is produced by immunizing a laboratory animal, often a mouse, with an antigen of interest. B cells, producing the specific antibody of interest, are selected and fused with an immortal myeloma cell line. The resulting fusion cell line, a hybridoma, becomes a factory for antibody production: antibody with specificity for one single epitope of the protein of interest. The resulting monoclonal "antiserum" is then produced inexpensively, in large quantities.

In immunodetection of proteins transferred during a Western blot, a primary (1°) antibody, the first antibody used, is selected for its specificity to the protein of interest. The primary antibody will react only with the protein of interest, allow-

Figure 19.1 Overview of Western blotting.

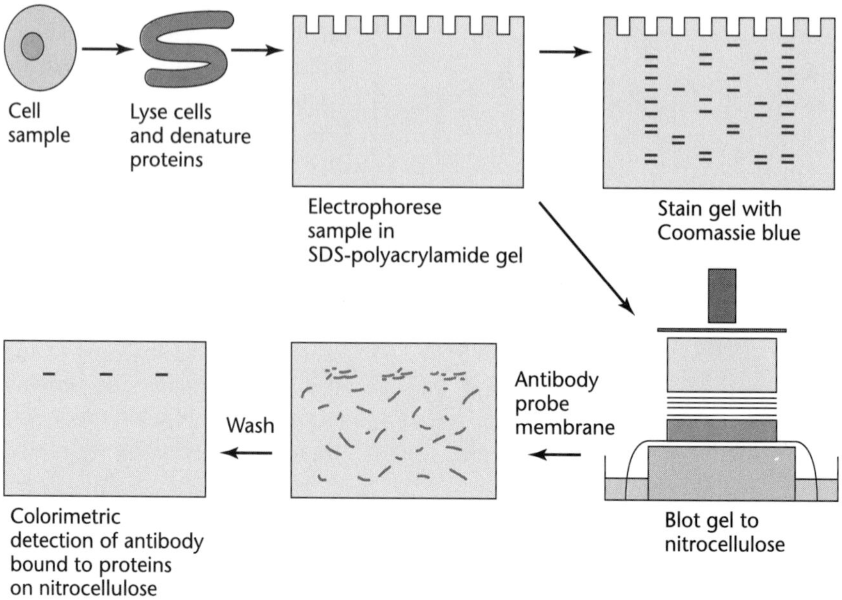

ing that protein to be picked out from all other proteins synthesized by a cell. The antibody may be conformationally dependent, also. In other words, the antibody will only recognize the protein of interest if it is folded correctly into its functional form.

For the binding of antibody to antigen to be visible, a chemical tag is required. The tag is a reagent that releases light, catalyzes a chemical reaction releasing light, catalyzes a precipitation reaction, or emits radioactivity. If there is no antigen-antibody interaction, no light or precipitation or radioactivity is observed. This reaction permits identification of a specific protein adhering to the Western blot nitrocellulose. The protein of interest is visualized separately from all other proteins present in a cell.

Sometimes this tag is conjugated directly to the primary antibody. Alternatively, the tag may be bound to a second antibody, one specific for the animal species in which the primary antibody was produced. This antibody is called the secondary (2°) antibody, since it binds to the primary antibody and is the second antibody used. The secondary antibody is used in detection to show that the primary antibody has bound to its antigen. In a direct immunoassay, the tag is placed on the primary antibody. For an indirect immunoassay, the tag is placed on the secondary antibody (Fig. 19.2). Regardless of its position, it is the tag, whether on the primary antibody or the secondary antibody, that shows that a specific antigen-antibody interaction has occurred.

Secondary antibodies are used for two reasons. First, it is expensive to tag each unique primary antibody. Secondary antibodies are specific for a species of antibody, such as mouse, rat, rabbit, or human. One may purchase an anti-rabbit antibody that will react to any primary antibody obtained from rabbits. This eliminates the need for tagging each unique primary antibody. The second reason to use a tagged secondary antibody is that the signal, whether a precipitate or radioactive signal, may be more intense. Secondary antibodies may bind multiple times to the primary antibody, thus amplifying the primary signal. However, using a secondary antibody does require more steps and more time to perform the experimental procedure.

Alkaline phosphatase is one commonly used antibody tag. The substrate for alkaline phosphatase is 5-bromo-4-chloro-3-indolylphosphate (BCIP). When BCIP is hydrolyzed, it causes oxidation of nitroblue tetrazolium (NBT), also included in the substrate reaction mixture, which then results in a blue-purple precipitate. Observation of the blue-purple precipitate is a positive reaction.

Applications of Western Blotting

Western blot analysis is performed when a researcher wants to determine whether a particular type of cell expresses, and therefore uses, a particular protein. DNA from cell type to cell type is the same. Cell types are differentiated from one another by the proteins they express. For example, islet cells of the pancreas produce insulin, used to regulate glucose metabolism. B cells from the immune system manufacture antibodies, used to fight off and control infectious diseases. Islet cells will never make antibodies, and B cells do not make insulin, even though each cell type contains the genes for insulin and antibody. A Western blot of proteins from islet cells is positive for insulin, not antibody, and a Western blot of proteins from a B cell is positive for antibody, not insulin.

Another example of protein expression differences occurs during development. Key genes in differentiation and morphogenesis may be turned on and off spatially, while an embryo develops. This provides a mechanism for correct production of a limb, brain, eye, or other organ in the correct location and size in a body. For example, in *Drosophila melanogaster*, the protein expressed from the maternal gene *bicoid* exists in a cytoplasmic gradient in the egg and determines the head-tail axis of the embryo. The greater concentration of bicoid is found at the head end. The bicoid protein activates at least one other gene, *hunchback*, whose protein product functions in establishing body segments. The orderly

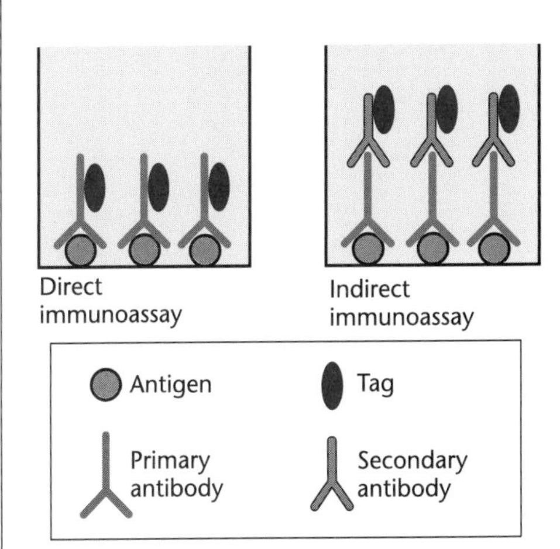

Figure 19.2 A direct immunoassay utilizes a tag on the primary antibody, the antibody that binds to the antigen. An indirect immunoassay utilizes a tag on the secondary antibody, which binds to the primary antibody.

progress of expression of one protein, depending on previous expression of a different protein, can be studied with Western blotting.

Western blotting may be used to diagnose the presence of a specific etiologic agent in an infectious disease. Infection by a bacterium or virus may be diagnosed by the presence of antibody specific for that agent. This antibody assay may be against all proteins from a given bacterium or virus. However, if just one protein from the infectious agent contains similarities to a second foreign protein to which an individual has been exposed, a false-positive reaction may occur. The patient may have antibodies to the second protein that cross-react with antigen from the etiologic agent, yielding a positive result when no infectious agent is present. Separating proteins based on size in an SDS-polyacrylamide gel provides a mechanism for assessing antibodies to a specific size of protein characteristic of the infectious agent.

The Western blotting technique may be used to analyze bacterially expressed proteins. Foreign genes are cloned into special plasmid expression vectors. Subsequently, the plasmids are transformed into bacteria. Because the expression vector possesses suitable promoter sequences, the bacteria can be forced to synthesize and express the protein coded for by the foreign gene. This provides researchers with ways to study protein function. Scientists may make mutations in a gene, or truncate the gene to cut off one end of a protein to determine which portions of the protein have a specific function. Different regions of the same protein may contribute to proper folding. Other regions may permit binding of one protein to another protein or to its receptor molecule. Sometimes, a point mutation is sufficient to alter the amino acid sequence of the protein, resulting in a nonfunctional molecule. Protein differences such as these are detectable by Western blotting.

Another extremely important use of protein expression from bacteria or other microorganisms involves the isolation and purification of a particular protein. Bacteria can be made to produce a desired protein, such as insulin, in great abundance relative to its own proteins. In this way, insulin can be mass produced more economically and safely than by extraction from porcine or bovine pancreas. Other proteins produced by this method include restriction enzymes, DNA polymerase, DNA ligase, and many other enzymes and proteins used both in molecular biology and in therapeutics. Western blotting can assist with ascertaining protein quantity and quality.

Ovalbumin

Albumins are proteins that are soluble in both water and salt solutions and are widely distributed in vertebrates. Albumins include lactalbumin in milk, serum albumin in blood, and ovalbumin in egg white.

Serum albumin is a major circulatory protein in many species, including humans, cows, and chickens. Found in plasma, it contributes to normal water balance between blood and tissues by regulating osmotic blood pressure. Albumin binds to a diverse set of ligands, contributing to their physiological distribution, transport, and metabolism into tissues. For example, the essential trace element zinc is bound to albumin for transport in blood, especially in cycling zinc between liver and pancreas.

A chicken egg, overall, contains about 11.8% total protein. Both egg white and egg yolk consist of a variety of proteins, but ovalbumin (molecular mass, about 45 kDa) accounts for approximately 54% of egg total protein. Ovalbumin is the most abundant protein in egg white. Egg yolk does not contain ovalbumin.

One might surmise that albumins, by virtue of their osmoregulatory properties, evolved to assist in the successful adaptation of vertebrates to terrestrial habitats. Ovalbumin clearly would have been an essential adjunct in preventing desiccation in reptile and bird eggs.

In this experiment, egg yolk, egg white, purified ovalbumin, and bovine serum albumin (BSA) (molecular mass, 66 kDa) will be used to illustrate both protein variety and the antigenic specificity of ovalbumin.

Papain

Papaya, or pawpaw (*Carica papaya*, family *Caricaceae*), is a smooth-textured musky-flavored fruit native to Central America. The *Carica* tree is a soft-wooded tree with flowers (and fruit) borne along the stems in leaf axils. By the time the fruits are mature, the leaves have been shed, leaving clusters of the ovoid fruits emerging directly from the stem. The demand for papaya in most of the world is not for their fresh fruit but for their production of papain, a protease enzyme extracted from the latex exuded by the skin of a green fruit. The fruits are tapped for latex when they are still immature and of a small size. Though these fruits may be tapped several times, once they mature they can be eaten. Papain, a protease extracted from the latex of *C. papaya* fruit, is characterized by the ability to gently hydrolyze proteins into smaller peptides, thus reducing a high-molecular-weight protein to lower-molecular-weight peptides or to amino acids.

Papain is a single polypeptide chain of 211 residues that consists of an alpha-helix and four strands of antiparallel beta-sheets. Papain contains the catalytic triad Cys-His-Asn, which categorizes papain as a cysteine proteinase. Its molecular mass is approximately 27,000 Da. Within its molecular structure are seven subsites, each capable of accommodating a single amino acid residue of a peptide substrate.

Because of its proteolytic effect, papain has wide commercial uses. It is most commonly used throughout the world as a meat tenderizer. Other uses include coagulating milk and clarifying beer. It increases the palatability of pet food. As a detergent, it removes proteinaceous stains. It also has several medical uses, including treatment of insect stings and other inflammatory processes and the acceleration of wound healing. In the leather industry, it is used to remove the hair from animal hides before tanning. In the last few years, papain has been used as an ingredient in cleaning fluids for soft contact lenses. In very low concentrations, papain is used in indigestion remedies and in the cosmetic industry as a mild skin exfoliant. In the research laboratory, papain has become a good tool for cell isolation since it can disrupt the extracellular matrix. Photoreceptor cells, cartilage cells, and neurons have been isolated in this manner.

Lab Language

antibody
antigen
conformational determinant
direct immunoassay
epitope
indirect immunoassay
monoclonal
primary antibody
polyclonal
secondary antibody

Laboratory Safety

Review the sections on physical, biological, and chemical hazards in "Laboratory Safety." Special considerations for this exercise:

boiling samples
SDS-PAGE: acrylamide, ammonium persulfate (APS), dithiothreitol (DTT), *N,N,N′,N′*-tetramethylethylenediamine (TEMED)
gel electrophoresis
NBT, BCIP

Laboratory Sequence

Two different protocols are provided for preparing proteins for Western blotting. Protocol 2A utilizes samples from egg, egg yolk, and egg white, for blotting to nitrocellulose. The blot is then probed with an antiovalbumin antibody. Protocol 2B describes how to produce lysates from papaya: seed, leaf, and fruit skin. Once blotted, these samples are probed with an antipapain antibody. It is intended that a particular class will prepare one set of samples (ovalbumin) or the other (papain) but not both at the same time. These protocols provide diversity in experimentation from one year to another and allow instructors to choose the sample preparation that best meets the needs of their students.

Day 1: Prepare gel and samples (gel may be prepared in advance and refrigerated).
Load gel and electrophorese.
Stain and destain gel.
Assemble Western blot and blot overnight.

Day 2: Dry the polyacrylamide gel.
Detect proteins bound to membrane using antibodies (incubations may go overnight).

Materials

SDS-PAGE

Class equipment and supplies
30% acrylamide-bisacrylamide, 37.5:1
1.5 M Tris, pH 8.8
10% SDS
10% APS
TEMED
1.0 M Tris, pH 6.8
Pasteur pipettes and bulbs
clinical centrifuge
microcentrifuge
heat block or boiling water bath
2× Laemmli sample buffer (LSB)
protein marker in 2× LSB
10× SDS running buffer
Coomassie blue stain
destain
gel-drying apparatus
1× physiologically buffered saline (PBS)
dH_2O

For ovalbumin:
chicken egg
1:10 egg yolk

Materials per pair of students
vertical gel electrophoresis system with plates, spacers, comb, and chamber gloves
microcentrifuge tubes
1 aliquot of 2× LSB
0.5- to 10-µl micropipette and sterile tips
10- to 100-µl micropipette and sterile tips
100- to 1,000-µl micropipette and sterile tips
power supply capable of 250 volts (shared by 2 to 4 groups)
500-ml graduated cylinder
50-ml graduated cylinder

Students will need to prepare
sample dilutions
10 or 15% polyacrylamide gel
5% stacking gel
300-ml 1× SDS running buffer

For ovalbumin:
1:10 egg white
10-mg/ml ovalbumin
10-mg/ml BSA

For papain:
papaya
papaya seeds
papaya leaf
fruit skin
meat tenderizer
0.5-µg/µl papain

Western Blot

Class equipment and supplies
10× Tris-glycine
methanol
blotting paper or paper towels
Whatman 3MM paper

Materials per pair of students
tile support and container for blotting
gloves

Students will need to prepare
300 ml of blotting buffer
3 wicks approximately 10 by 30 cm
nitrocellulose cut to gel size
3 layers of Whatman paper cut to gel size

Immunodetection

Class equipment and supplies
NaCl
KCl
Na_2HPO_4
KH_2PO_4
Tween 20
dried milk
primary antibody (1° Ab)
- antiovalbumin
- antipapain

secondary antibody (2° Ab) conjugated to alkaline phosphatase
25× developer concentrate
NBT solution
BCIP solution
shaking platform (optional)

Materials per pair of students
0.5- to 10-μl micropipette and sterile tips
10- to 100-μl micropipette and sterile tips
container for incubations

Students will need to prepare
500-ml PBS with 0.5% Tween 20 (PBS-T)
25-ml PBS-T with 5% dried milk
15-ml 1:1,000 1° Ab in PBS-T
15-ml 1:1,000 2° Ab in PBS-T
25-ml developer solution

Protocols

PROTOCOL 1 PREPARATION OF THE SDS-POLYACRYLAMIDE GEL

1. **Wearing gloves,** clean the glass plates with soap and water and rinse well.

Clean plates help prevent bubble formation when acrylamide gels are poured.

2. Assemble the gel apparatus (Fig. 19.3). The short glass plate should face outward and the longer plate inward. The clamp knobs should face inward.

Align the glass plates and spacers on a flat surface.

3. Pipette water between the glass plates to check that the apparatus has been assembled correctly so the acrylamide solution will not leak out when the gel is poured. Pour the water out and blot with a paper towel.

The glass plates and spacers should form a watertight seal against the gasket.

4. **Wearing gloves**, mix the reagents together for 5 ml of 10% (ovalbumin) or 15% (papain) polyacrylamide resolving gel in a disposable tube. Add reagents in the order listed in the table below.

*Acrylamide is a neurotoxin: **avoid skin contact.***

APS initiates polymerization.

TEMED speeds up the polymerization process.

Reagent	10% Gel	15% Gel
dH_2O	2.00 ml	1.15 ml
1.5 M Tris, pH 8.8	1.25 ml	1.25 ml
30% acrylamide-bisacrylamide	1.70 ml	2.50 ml
10% SDS	50 μl	50 μl
10% APS	50 μl	50 μl
TEMED	5 μl	5 μl

5. Immediately "pour" the gel using a Pasteur pipette. Hold the comb outside the glass plates and fill the plates to about 5 mm below the lowest level of the comb.

6. Make sure there are no bubbles in the gel. Sharply tap the glass to force any bubbles to the top of the gel.

7. Add a thin layer of isopropanol or isobutanol to the top of the gel.

Polymerization is inhibited by exposure to air.

8. Keep the tube with the acrylamide solution. When the solution has solidified in the tube, this indicates that the gel has solidified too, and the stacking gel can be poured.

Begin preparation of your samples while the gel is solidifying.

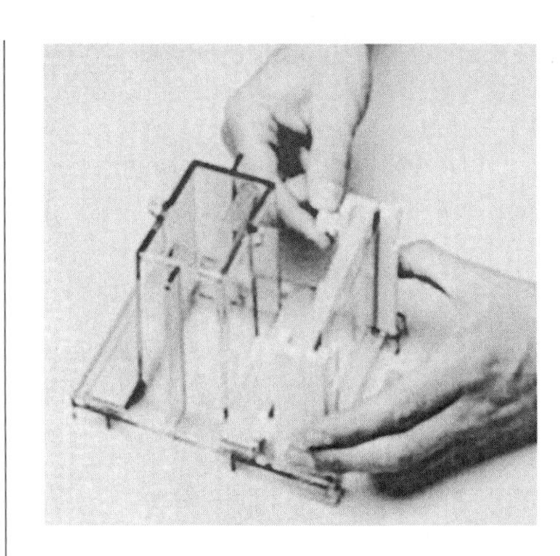

Figure **19.3** Assembly of gel apparatus. (Reprinted from Bio-Rad Laboratories, Hercules, Calif., with permission.)

9. Decant the alcohol from the 10 or 15% polyacrylamide gel and rinse several times with dH_2O to remove all the alcohol.

The stacking gel ensures that all proteins enter the resolving gel together.

10. **Wearing gloves**, prepare 2.5 ml of the 5% stacking gel in a disposable tube. Add the reagents in the order listed in the table below.

Reagent	Volume
dH_2O	1.70 ml
1.0 M Tris, pH 6.8	313 µl
30% acrylamide-bisacrylamide	416 µl
10% SDS	25 µl
10% APS	13 µl
TEMED	2.5 µl

11. Pour the stacking gel between the glass plates using a Pasteur pipette. Insert the comb between the glass plates, taking care not to get bubbles stuck underneath the wells.

12. Keep the extra stack solution to assess when the stacking gel has solidified.

PROTOCOL 2A PREPARATION OF EGG SAMPLES FOR WESTERN BLOTTING

1. Obtain the following stock samples: 10-mg/ml ovalbumin (ova), 10-mg/ml BSA, 1:10 egg yolk, and 1:10 egg white.

2. Prepare 1:10 dilutions of the ova and the BSA by mixing 100 µl of each of the 10-mg/ml stock solutions with 900 µl of PBS.

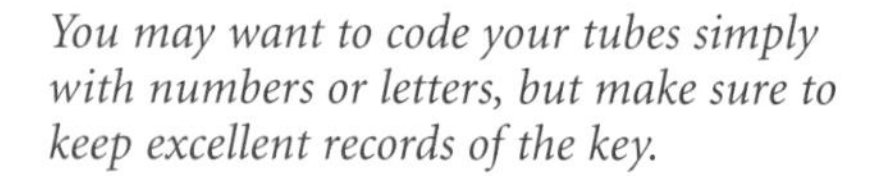

You may want to code your tubes simply with numbers or letters, but make sure to keep excellent records of the key.

3. Prepare 1:100 and 1:1,000 dilutions of egg yolk. Mix 100 µl of the 1:10 dilution of egg yolk with 900 µl of PBS. Mix 100 µl of the 1:100 dilution of egg yolk with 900 µl of PBS. Label the tubes clearly.

4. Prepare 1:100 and 1:1,000 dilutions of egg white . Mix 100 µl of the 1:10 dilution of egg white with 900 µl of PBS. Mix 100 µl of the 1:100 dilution of egg white with 900 µl of PBS. Label the tubes clearly.

2× LSB contains SDS, which gives all proteins a net negative charge, and dithiothreitol, which breaks disulfide bonds.

5. Label four clean microcentrifuge tubes. Mix 25 µl of each *diluted sample* (ova, BSA, 1:1,000 egg yolk, and 1:1,000 egg white) with 25 µl of 2× LSB.

The protein marker contains proteins of known sizes for comparison with unknown samples.

6. Obtain a protein marker in 2× LSB.

7. Heat all of the samples and the protein marker for 5 minutes at 100°C, *just before loading the gel.*

8. Load 10 µl of each sample to the gel as diagrammed in step 5 of protocol 3.

9. Calculate the amount of ova and BSA loaded to a gel lane. Remember to account for the dilution plus mixing each sample 1:2 with 2× LSB and the quantity loaded to the gel.

EXERCISE 19

Heating assists with the protein denaturation.

LSB also contains glycerol, to make the sample heavy for loading purposes, and bromophenol blue, to monitor the progress of electrophoresis.

PROTOCOL 2B PREPARATION OF PAPAYA SAMPLES FOR WESTERN BLOTTING

A. Preparation of Papaya Seed Lysate

1. Obtain four frozen papaya seeds.

Frozen tissue is easier to dissociate.

2. Using a chilled mortar and pestle, grind the seeds into a paste.

3. Transfer the paste to a 1.5-ml microcentrifuge tube.

4. Add 500 µl of 2× LSB.

5. Vortex to mix and lyse the cells.

6. Incubate the sample at room temperature for 5 minutes, flicking occasionally to mix.

7. Centrifuge at 14,000 rpm for 3 minutes to pellet the tissue remnants.

Chunks of debris will interfere with electrophoresis.

8. Remove the supernatant to a clean, labeled microcentrifuge tube and use for SDS-PAGE analysis.

9. Prepare a 1:2 dilution of the lysate.

10. Heat the samples at 100°C for 5 minutes, *just before loading the gel.*

B. Preparation of Leaf Lysate

1. Collect a fresh, tender leaf. Cut a 1-cm-square piece of tissue from it.

2. Place the leaf tissue in a microcentrifuge tube. Crush it into a paste with a toothpick or small pestle.

Freezing tissue will make it easier to crush.

3. Add 100 µl of 2× LSB and vortex to mix.

4. Incubate the tube at room temperature for 5 minutes, flicking occasionally to mix.

5. Microcentrifuge at 10,000 rpm for 3 minutes to separate particulate material from leaf protein extract.

Chunks of leaf debris will interfere with electrophoresis.

6. Using a pipette, transfer the supernatant into a clean, labeled microcentrifuge tube.

7. Prepare a 1:2 dilution of the lysate.

8. Heat all samples, at 100°C for 5 minutes, *just before loading the gel.*

C. Preparation of Fruit Skin Lysate

1. Obtain a 3- by 1-cm rectangle of frozen papaya skin from the fruit.

The sand will assist with dissociating skin cells.

2. Place the fruit skin in a chilled mortar and add a pinch of sea sand.

3. Add 100 µl of 2× LSB.

4. Grind the fruit, sand, and LSB to a paste with the chilled pestle.

5. Using a small spatula (or gloved finger), transfer the paste into a microcentrifuge tube.

6. Incubate the tube at room temperature for 5 minutes.

Sand and particulate matter will interfere with electrophoresis.

7. Centrifuge at 10,000 rpm for 3 minutes to pellet particulate material from the fruit skin lysate.

8. Using a pipette, transfer the supernatant into a clean, labeled microcentrifuge tube.

9. Prepare a 1:2 dilution of the lysate.

10. Heat all samples at 100°C for 5 minutes, *just before loading the gel.*

D. Preparation of Meat Tenderizer

1. Weigh two portions of meat tenderizer, 0.5 g each, into two microcentrifuge tubes.

2. Add 500 µl of 2× LSB to one tube.

3. Vortex vigorously to mix.

4. Microcentrifuge at 14,000 rpm for 2 minutes.

5. Remove the supernatant with a pipette and add to the second tube of meat tenderizer.

6. Vortex vigorously to mix.

7. Microcentrifuge at 14,000 rpm for 2 minutes.

8. Remove the supernatant to a clean, labeled microcentrifuge tube.

9. Heat all samples at 100°C for 5 minutes, *just before loading the gel.*

PROTOCOL 3 SDS-PAGE

1. Transfer your gel to the electrophoresis chamber. Two groups will share an electrophoresis chamber to electrophorese two gels.

2. Prepare 300 ml of 1× SDS running buffer from a 10× stock solution. One gel chamber requires 300 ml.

3. Add 1× SDS running buffer to the inside chamber so it covers the wells.

4. Pour running buffer into the bottom of the chamber until it makes contact with the bottom of the gel.

5. Gently remove the comb.

6. Load 10 µl of sample to each well. For the ovalbumin blot, set up your gel in duplicate.

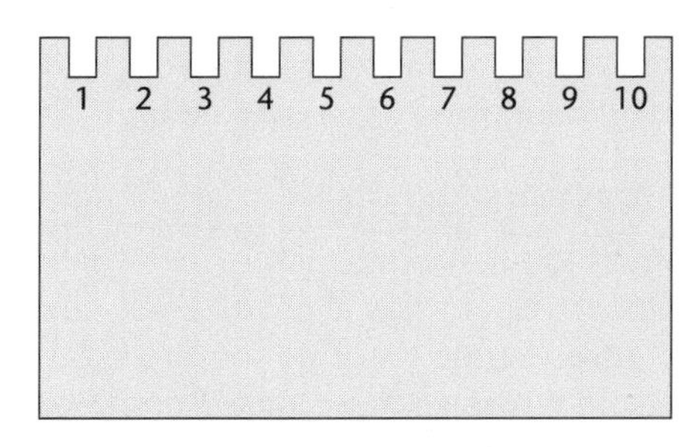

Lane	Ovalbumin	Papain
1	Protein marker	Protein marker
2	Ovalbumin	Papain
3	BSA	Meat tenderizer
4	Egg yolk	Fruit lysate
5	Egg white	Fruit 1:2 lysate
6	Protein marker	Leaf lysate
7	Ovalbumin	Leaf 1:2 lysate
8	BSA	Seed lysate
9	Egg yolk	Seed 1:2 lysate
10	Egg white	Protein marker

7. Place the cover on the gel apparatus and plug the electrodes into the power supply.

8. Turn on the power supply and electrophorese at 200 volts for about 1 hour, until the bromophenol blue dye front reaches the bottom of the gel.

PROTOCOL 4 COOMASSIE BLUE STAINING AND DESTAINING

When the electrophoresis is complete, **turn off the power supply. Wearing gloves,** carefully take the apparatus apart.

1. Remove the gel apparatus from the buffer chamber and remove the gel in the plates from the holder.

2. Remove the spacers and use one of the spacers to gently pry the glass plates apart. Take care not to rip the gel.

3. Cut off a small corner of the gel at the lower edge of the marker lane for orientation.

Ovalbumin blot:

4. Cut the acrylamide gel in half between lanes 5 and 6, using the marker lane as a guide.

5. Stain one-half of the gel in Coomassie blue by incubating at room temperature for 30 minutes. Use the remaining half of the gel for Western blotting.

Papain blot:

4. One pair of students will stain their gel in Coomassie blue by incubating at room temperature for 30 minutes.

One-half of the ovalbumin gel will be stained with Coomassie blue, and one-half of the gel will be blotted.

Prepare the blotting materials—wicks, membrane, Whatman paper, blotting buffer, and 500 ml of PBS-T—while electrophoresing.

A fine stream of bubbles from the electrodes indicates current flow.

The glass plates are very fragile.

Moisten gloves and gel with water contained in the squirt bottle.

Coomassie blue stains all proteins present in the acrylamide gel.

Gels may be stained and destained for shorter times at 60°C.

5. The other pair of students using the gel chamber will blot their gel to nitrocellulose (protocol 5).

Coomassie blue-stained gel:

The Kimwipe scavenges stain out of the solution and binds it.

6. Destain the gel by incubating in destain solution 30 minutes to overnight at room temperature. Crumple a Kimwipe and place it in the corner of the container, in the destain solution, to absorb the stain.

PROTOCOL 5 WESTERN BLOT

To avoid contaminating the membrane with oils, do not touch the membrane with ungloved fingers.

1. Prepare the materials for the Western blot.
   - Cut three Whatman paper wicks about 10 by 30 cm.
   - Cut a piece of nitrocellulose the same size as the gel.
   - Cut three pieces of Whatman paper the same size as the gel.
   - Cut 4 inches of paper towels the same size as the gel.

SDS-polyacrylamide gels may be electroblotted or blotted using a semidry system.

2. Prepare blotting buffer. Determine the amount of each stock solution to use to make 300 ml.

Reagent	Stock concentration	Final concentration	Amount of stock to make 300 ml
Tris-glycine	10×	1×	
Methanol	100%	20%	

Add the Tris-glycine and methanol to dH_2O in a graduated cylinder. Adjust the volume to 300 ml with dH_2O.

3. **Wearing gloves**, pre-wet the nitrocellulose membrane with dH_2O and then soak in blotting buffer.

Bubbles in between layers will impede even and efficient protein transfer.

4. Use the unstained half of the gel to set up the Western blot, as shown in Fig. 19.4. Take special care to avoid any air bubbles between the layers.

5. Wet the wicks and drape them over the tile support so they make contact with the buffer at each end.

The wick draws buffer through the gel, transferring the DNA from the gel to the membrane.

6. Carefully place the gel on the wicks and cover with the nitrocellulose membrane and three layers of wet Whatman paper.

7. Use a pipette like a rolling pin to roll bubbles out from under the Whatman paper and membrane before the paper towel layer is added.

The paper towels draw buffer through the gel via capillary action to transfer the proteins onto the membrane.

8. Place about 4 inches of dry paper towels on top of the Whatman paper with a weight on top.

9. Allow the proteins to transfer overnight.

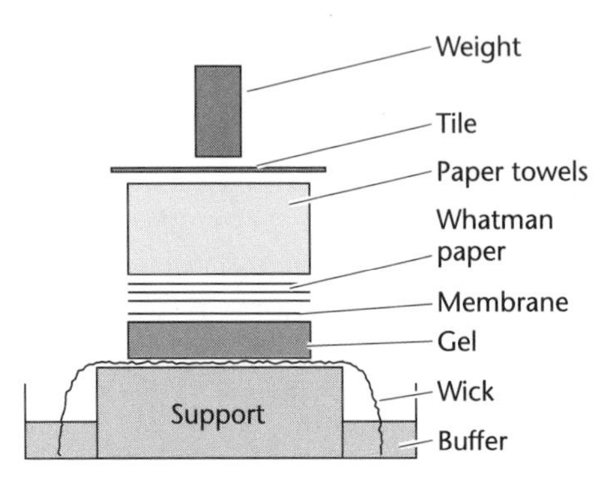

Figure 19.4 Assembly of the Western blot.

PROTOCOL 6 DRYING THE POLYACRYLAMIDE GEL

Once the Coomassie blue-stained portion of the gel has been sufficiently destained, it may be air dried between pieces of gel-drying film as a permanent record of the gel electrophoresis.

1. Wet two pieces of gel-drying film by soaking in water.
2. Place one piece of film on top of one of the gel frames.
3. Gently place the acrylamide gel on the film. Use a squirt bottle with water to remove bubbles from between the gel and the film and to keep the gel moist.
4. Carefully sandwich the gel between the bottom and top pieces of the film. *Make sure all of the air bubbles get pushed out from between the two layers of film.*
5. Align the second gel frame on top of the first and securely clamp the two layers of film with the gel between the two frames.
6. Allow the gel to air dry overnight.

PROTOCOL 7 DETECTION OF PROTEINS BOUND TO NITROCELLULOSE

1. Prepare PBS-T (PBS with 0.5% Tween 20). Determine the amounts of reagents to use to prepare 500 ml.

Tween 20 is a detergent.

See Fig. 19.5.

Reagent	Concentration for 1×	Amount for 1 liter	Amount for 500 ml
NaCl	137 mM	8.0 g	
KCl	2.7 mM	0.2 g	
$Na_2HPO_4 \cdot 7H_2O$	4.3 mM	1.15 g	
KH_2PO_4	1.4 mM	0.2 g	
Tween 20	0.5%	5.0 ml	

Add the powdered reagents to dH_2O in a graduated cylinder while stirring. Adjust the volume to 500 ml when all components are dissolved.

2. **Wearing gloves,** carefully take the blot apart. Use a razor blade to cut off a small piece of the lower corner of the nitrocellulose at the marker lane.
3. Incubate the nitrocellulose membrane in PBS-T with 5% blocking reagent (powdered milk) for 30 to 60 minutes.
4. Rinse the membrane two times in PBS-T.
5. Wash the membrane by incubating it in PBS-T for 10 minutes at room temperature.
6. Discard the wash and add the primary (1°) antibody diluted in PBS-T. Incubate for 1 hour at room temperature with occasional shaking.
7. Remove the 1° antibody and save at 4°C for future use.
8. Rinse the membrane two times in PBS-T.
9. Wash the membrane two times, for 10 minutes each time, in PBS-T.
10. Discard the wash and add the secondary (2°) antibody diluted in PBS-T.
11. Incubate the membrane for 30 minutes at room temperature.

The blocking reagent is usually a protein solution, such as powdered milk, that serves to bind or block sites on the nitrocellulose where protein from the gel has not bound.

Blocking may be done at 4°C overnight. The membranes may be kept for long periods of times if stored moist at 4°C.

Change the blocking reagent every 2 or 3 days if the membrane is stored longer than overnight.

To rinse means to add buffer and immediately pour it off.

To wash means to add buffer and allow the membrane to incubate in buffer for the specified period.

The 2° antibody is tagged with alkaline phosphatase.

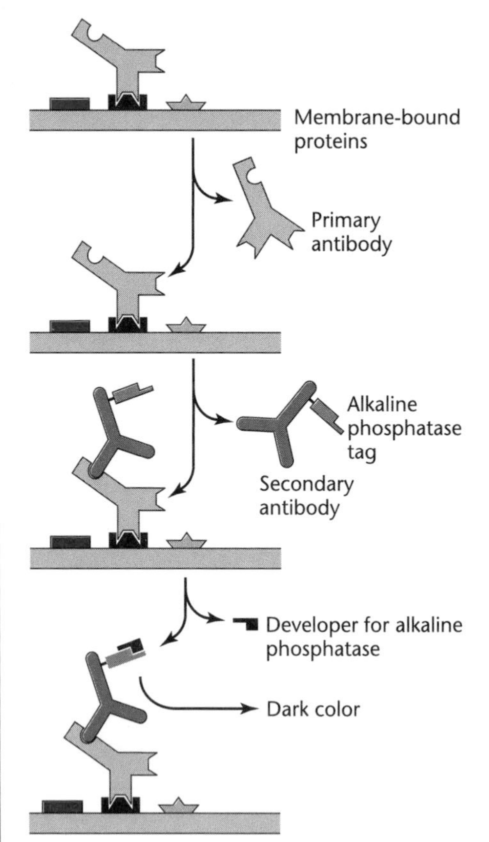

Figure 19.5 Indirect immunodetection of proteins after Western blotting.

12. Rinse the membrane two times in PBS-T.

13. Wash the membrane two times, for 10 minutes each time, at room temperature.

14. Prepare developer solution immediately before use, and protect from light exposure. Mix 24 ml of dH_2O with 1 ml of 25× buffer concentrate, 0.25 ml of NBT solution, and 0.25 ml of BCIP solution.

Alkaline phosphatase hydrolyzes BCIP, which then oxidizes NBT to form a blue-purple precipitate.

15. Pour off the last wash, and add the developer solution to the membrane. Incubate in the dark. Color will develop in 5 minutes to overnight.

16. Stop the reaction by rinsing the membrane with dH_2O.

References and General Reading

Ausubel, F. M., R. Brent, R. E. Kingston, D. D. Moore, J. G. Seidman, J. A. Smith, and K. Struhl (ed.). 1994. *Current Protocols in Molecular Biology.* Wiley Interscience, New York, N.Y.

Belitz, H. D., and W. Grosch. 1987. *Food Chemistry.* Springer-Verlag, Heidelberg, Germany.

Brocklehurst, K., E. Salih, R. McKee, and H. Smith. 1985. Fresh non-fruit latex of *Carica papaya* contains papain, multiple forms of chymopapain A and papaya proteinase omega. *Biochem. J.* **228:**525–527.

Kamphuis, I. G., K. H. Kalk, M. B. Swarte, and J. Drenth. 1984. Structure of papain refined at 1.65 Å resolution. *J. Mol. Biol.* **179:**233–256.

Silva, L. G., O. Garcia, M. T. Lopes, and C. E. Salas. 1997. Changes in protein profile during coagulation of latex from *Carica papaya. Braz. J. Med. Biol. Res.* **30:**615–619.

Simpson, B. B., and M. C. Ogorzaly. 1995. *Economic Botany: Plants in Our World,* 2nd ed. McGraw-Hill Book Co., New York, N.Y.

Spector, W. S. (ed.). 1956. *The Handbook of Biological Data.* The W. B. Saunders Co., Philadelphia, Pa.

Whitney, E. N., C. B. Cataldo, and S. R. Rolfes. 1998. *Understanding Normal and Clinical Nutrition,* 5th ed. West/Wadsworth, Belmont, Calif.

Worthington, V. (ed.). 1993. *Worthington Enzymes and Related Biochemicals Manual.* Worthington Biochemical Corporation, Lakewood, N.J.

Zpalis, C., and A. Beck. 1985. *Food Chemistry and Nutritional Biochemistry.* John Wiley & Sons, Inc., New York, N.Y.

Interesting Internet Sites
http://www.rain-tree.com/papaya.htm (provides botanical information)
http://www.valleyenzymes.com (Valley Research Inc.)
http://www.yorvic.york.ac.uk (provides protein information)

Discussion

1. What purpose does Coomassie blue staining serve?
2. Describe how blotting is accomplished in the Western blotting technique.
3. Why are proteins transferred to and immobilized onto a membrane?
4. What do you observe on the nitrocellulose membrane? Explain.
5. How will the nitrocellulose membrane be analyzed?
6. What is the purpose of a primary antibody?
7. What is the purpose of a secondary antibody?
8. What is the purpose of the blocking step?

Observe your Coomassie blue-stained gel.

9. Locate the protein marker lane of the Western blot. What do you observe? What is the marker lane and what is its purpose?
10. Observe your experimental sample lanes. Do you see any differences? Do you see any similarities? Why do you think this is so?

Observe the antibody-stained membrane.

11. Locate your marker lane. What do you observe? Why do you think this is so?
12. Locate your experimental lanes. What do you observe? Are there any differences or similarities between the experimental lanes? Why do you think this is so?
13. Western blotting is more specific than an ELISA (enzyme-linked immunoabsorbent assay) or other tests for proteins from a virus, such as human immunodeficiency virus type 1, when used for diagnosis. Why do you think this is so?

SECTION VI

Gene Transfer

EXERCISE 20

Transformation of Yeasts

Key Concepts

1. Yeasts are single-cell eukaryotic organisms that contain a relatively small genome yet display many cellular and molecular features in common with higher eukaryotes.
2. Yeasts are easy to propagate and their genome is easy to manipulate, qualities that make them a good model organism for studying eukaryotic processes and for producing eukaryotic proteins.

Goals

By the end of this exercise, you will be able to

- discuss the importance of using yeasts in biotechnology
- explain transformation in yeasts
- compare and contrast the uses of yeasts and bacteria

Introduction

Yeasts, like bacteria, are important in pharmaceutical and industrial applications of biotechnology. Both cell types have a high surface-to-volume ratio, which makes rapid nutrient uptake and biosynthesis possible. Both are genetically manipulable with facility. *Saccharomyces cerevisiae* can be transformed with plasmid eukaryotic expression vectors as easily as *Escherichia coli* can be transformed with bacterial plasmids. *S. cerevisiae* has a generation time of about 90 minutes and, like bacteria, can be grown both in liquid and on semisolid agar media. Culture conditions for bacteria and yeasts are very similar, although yeasts require a richer medium and a cooler incubation temperature of 25 to 30°C. Like bacteria, yeasts are also easily grown in small-batch fermentors as well as in large-scale bioreactors, requirements for commercially produced recombinant expression products.

These two organisms, yeasts and bacteria, are widely used in biotechnology because of their ease of manipulation and growth. However, they belong to two structurally distinct cell types: yeasts are eukaryotic, and bacteria are prokaryotic. These two cell types differ in their internal organization, in features that may make a prokaryotic organism better for use in a specific biotechnology application than a eukaryotic organism, or vice versa.

S. cerevisiae, also known as baker's yeast, is commonly used as a model organism for the study of more complex eukaryotes. An important difference between prokaryotes and eukaryotes is that eukaryotic proteins may need to undergo posttranslational modification to become functional. They may be cleaved by proteases, they may be glycosylated (sugars added to specific amino acids), or disulfide bridges may be formed. Recombinant protein expression systems in bacteria cannot carry out the processes necessary to produce functional proteins. *S. cerevisiae* serves well as a producer of authentic human recombinant proteins; it can perform many posttranslational modifications that prokaryotic host cells cannot. Recombinant human therapeutic proteins synthesized by *S. cerevisiae* include epidermal growth factor, blood coagulation factor XIIIa, and insulin. The use of *S. cerevisiae* is not subject to the same governmental regulations applied to many other host cells used in producing therapeutics for human consumption, because it has received approval by the U.S. Food and Drug Administration as an organism "generally recognized as safe." This is an outgrowth of the many years that *S. cerevisiae* has been used in home and commercial baking and brewing.

Transformation of yeasts with plasmid DNA is similar to the process of bacterial transformation: yeasts are made competent by means of a cation solution, DNA is introduced, heat shock triggers DNA uptake, and then the cells are plated on selective growth medium. Since yeasts are eukaryotic, the details of each step differ from those of bacteria. Yeasts are made competent by incubation in lithium acetate (LiAc), an alkali metal cation. Next, the plasmid DNA is introduced with a single-stranded carrier DNA and polyethylene glycol (PEG). The carrier DNA increases transformation efficiency by 1 or 2 orders of magnitude. The PEG treatment, coupled with heat shock, makes the cell wall permeable and triggers DNA uptake. Finally, the transformed cells are plated on selective medium to determine which yeast cells have successfully accepted the foreign DNA.

Yeast cells have been successfully transformed using electroporation or spheroplast transformation. Electroporation, as with bacteria, is extremely efficient, especially when only small quantities of plasmid DNA are available. It is also rapid but does require use of expensive electroporation equipment. Spheroplast transformation involves removal of the cell wall with enzymatic treatment. DNA is introduced into the resulting spheroplast with PEG. Spheroplasts must be handled gently and can be easily disrupted by vortexing, vigorous pipetting, and osmotic shock. Transformed spheroplasts must be plated in top agar, a thin layer of softer (lower-concentration) agar on top of the regular agar layer in the plate, to provide an environment for regeneration of the cell wall. Spheroplast transformation probably provides the greatest transformation efficiency but is the most tedious and time-consuming. It should also be noted that each strain of yeast will vary in its optimal conditions for electroporation or spheroplast transformation. The LiAc method is least affected by differences in yeast strains.

Strains and Plasmids for Gene Expression in *S. cerevisiae*

Wild-type yeasts are nutritionally self-sufficient (prototrophic), capable of growing on minimal media. Auxotrophic mutants, however, cannot grow on minimal media; these strains require specific nutritional supplements, which depend on the individual auxotroph. Auxotrophy provides a mechanism for selecting successfully transformed yeast cells, those containing one or more recombinant plasmids.

For example, in this experiment, we will use *S. cerevisiae* YPH499. It is an auxotrophic mutant with an impaired *TRP1* gene (which encodes an enzyme in the synthesis of tryptophan) and an impaired *URA3* gene (which encodes an enzyme in the uracil biosynthesis pathway). This strain requires exogenous tryptophan and uracil in the growth medium, since it cannot synthesize these two essential metabolites. However, upon transformation with a plasmid (or plasmids) containing a normal copy of the *TRP1* and *URA3* genes, the nutritional requirements of strain YPH499 will be satisfied: it will possess the capability of performing its own biosyntheses and growing on minimal media. This strain can use galactose as a carbon source.

In addition to selectable markers, yeast expression vectors also must carry a eukaryotic promoter to regulate protein expression. Two such promoters are *GAL1* and *GAL10,* which are induced, or activated, in the presence of galactose. Finally, the yeast plasmid carries an expression termination sequence that provides proper termination of RNA transcripts.

In this laboratory experiment, we will transform *S. cerevisiae,* a suitable yeast host, with two plasmids. Together they bear the genes, taken from *Erwinia herbicola,* required for biosynthesis of beta-carotene. Our experimental protocols will enable us to determine not only successful transformation, but also partial or complete synthesis of the beta-carotene pathway. When beta-carotene is synthesized, yeast colonies will be pigmented yellow. In the absence of lycopene cyclase, lycopene, a red pigment, accumulates instead.

Carotenoids

Carotenoids are naturally occurring red, orange, and yellow pigments. They are found in all three domains of life: *Bacteria, Archaea,* and *Eucarya.* Approximately 300 different types of carotenoids have been found in living organisms, including lycopene, the red pigment of tomatoes; beta-carotene, the orange pigment of carrots; and zeaxanthin, the yellow pigment of corn. Various algae are rich sources of carotenoids from which more than 100 different types have been isolated and characterized.

In plants, algae, and photosynthetic bacteria, carotenoids act as antenna pigments in the light-harvesting process of photosynthesis by absorbing light at wavelengths other than those absorbed by chlorophyll. Carotenoids appear to protect against destructive photooxidation by ultraviolet light. They provide coloration in animals such as flamingos, ibises, canaries, goldfish, and salmon.

Carotenoids are important factors in human health. Beta-carotene, for instance, is the main dietary precursor of vitamin A, consumed in plant foods. Recent research indicates that carotenoids may protect against cancer and heart disease owing to their role as antioxidants. In addition, carotenoids are used as natural colorant additives in drinks and animal feed. The industrial production of natural carotenoids using microbial biotechnology is already an established and expanding practice.

Carotenoid Biosynthesis in *E. herbicola*

Erwinia is a gram-negative bacillus in the family *Enterobacteriaceae,* as is *E. coli.* Most members of the genus *Erwinia* are plant pathogens or contribute to vegetable spoilage. A few are saprophytes (decomposers). *E. herbicola* produces a yellow pigment characterized as beta-carotene, determined by a biochemical pathway involving four different enzymes, illustrated in Fig. 20.1. We will use this biochemical pathway in our transformation experiment.

The plasmids we will use in this experiment are pARC145G and pARC1520. Plasmid pARC145G contains *GAL10* and *GAL1* positioned as divergent promoters (Fig. 20.2). The geranylgeranyl pyrophosphate (GGPP) synthase gene is adjacent to the *GAL10* region, and the gene for phytoene synthase is located on the *GAL1* side. Thus, the GGPP synthase gene is expressed in *S. cerevisiae* using the *GAL10* promoter, and phytoene synthase expression is controlled by the *GAL1* promoter. pARC145G also carries the *TRP1* gene to permit selection of transformants, and the Amp^r gene and a bacterial origin of replication (*ori*) for selection and propagation in *E. coli.* Thus, pARC145G is a shuttle vector, capable of propagation in the prokaryote *E. coli* or the eukaryote *S. cerevisiae.*

In plasmid pARC1520, the gene for phytoene dehydrogenase-4H is adjacent to the *GAL10* region of the *GAL10-GAL1* divergent promoter; the gene for lycopene cyclase is located on the *GAL1* side. Phytoene dehydrogenase is expressed in *S. cerevisiae* using the *GAL10* promoter, and lycopene cyclase expression is controlled by the *GAL1* promoter. This plasmid also carries the *URA3* gene, which allows for selection of transformants. Like pARC145G, pARC1520 also carries the Amp^r gene and a bacterial *ori* for selection and propagation in *E. coli* and is a shuttle vector.

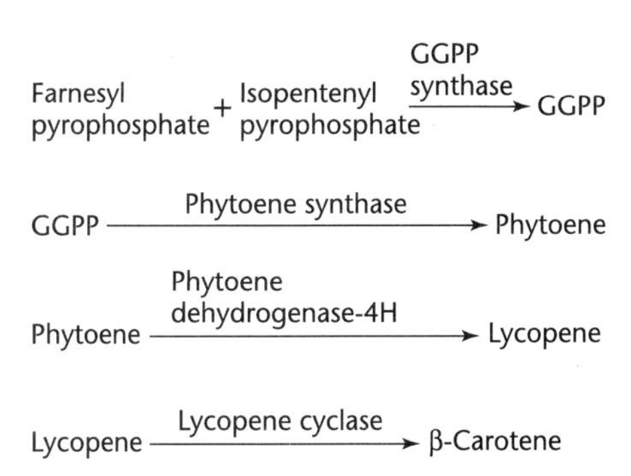

Figure 20.1 Synthesis of beta-carotene. Beta-carotene is synthesized from farnesyl pyrophosphate and isopentenyl pyrophosphate. This pathway requires four enzymes: geranylgeranyl pyrophosphate synthase, phytoene synthase, phytoene dehydrogenase-4H, and lycopene cyclase. When beta-carotene is synthesized, yeast colonies will be pigmented yellow. In the absence of lycopene cyclase, lycopene, a red pigment, accumulates instead.

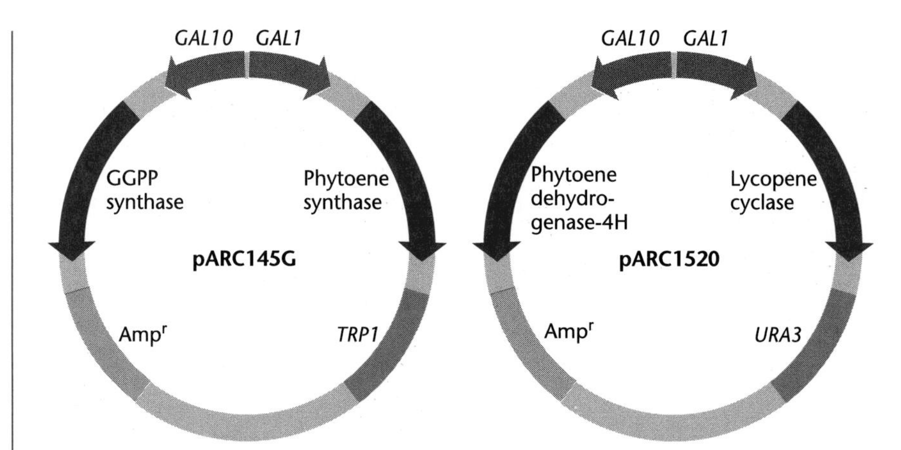

Figure 20.2 Plasmids pARC145G and pARC1520, used to transform *S. cerevisiae* for production of beta-carotene. pARC145G contains the genes for GGPP synthase and phytoene synthase. pARC1520 contains the genes for phytoene dehydrogenase-4H and lycopene cyclase. All four gene products must be present for synthesis of beta-carotene.

Lab Language

auxotrophic
beta-carotene
bioreactor
carotenoids
E. herbicola
GAL1
GAL10
GGPP synthase
glycosylation
lycopene
lycopene cyclase
minimal media
phytoene dehydrogenase-4H
phytoene synthase
posttranslational modification
prototrophic
shuttle vector
spheroplasts
TRP1
URA3

Laboratory Safety

Review the sections on physical, biological, and chemical hazards in "Laboratory Safety." Special considerations for this exercise:

hot autoclaved solutions
working with microbes

Materials

Class equipment and supplies

single-stranded carrier DNA (2 mg/ml)
1 M LiAc, pH 8.5
100 mM LiAc
50% PEG
yeast nitrogen base, without amino acids
10 M NaOH
20% galactose solution
10× dropout mixture
10× dropout mixture without uracil
10× dropout mixture without tryptophan
10× dropout mixture without uracil and tryptophan
synthetic complete (SC) media agar plates
SC − uracil agar plates
SC − tryptophan agar plates
SC − uracil and tryptophan agar plates
microcentrifuge
vortex
42°C water bath or heat block
30°C incubator

Materials per pair of students

wax or permanent marking pencil
0.1- to 10-μl micropipette and sterile tips

10- to 100-μl micropipette and sterile tips
100- to 1,000-μl micropipette and sterile tips
sterile microcentrifuge tubes
gloves
inoculating loop
Bunsen burner
1,000-ml beaker to collect biological waste
1 plate of overnight culture of *S. cerevisiae* YPH499 on SC agar
1 aliquot of 100 mM LiAc
1 aliquot of 1 M LiAc
1 aliquot of pARC145G (0.2 μg/μl)
1 aliquot of pARC1520 (0.2 μg/μl)
1 aliquot of 50% PEG solution
1 aliquot of single-stranded carrier DNA
1 aliquot of sterile dH_2O
1 SC plate
1 SC − uracil plate
1 SC − tryptophan plate
1 SC − uracil and tryptophan plate

Protocols

PROTOCOL 1 YEAST TRANSFORMATION

1. Obtain an overnight agar culture of *S. cerevisiae.*
2. Pipette 1.0 ml of sterile dH_2O into a sterile microcentrifuge tube.
3. Using a sterile inoculating loop, scrape a loopful of yeast from the plate and transfer it into the dH_2O. Collect cells from the edge of a colony.

Cells from the edge of the colony will be in log phase.

4. Resuspend the cells by vortexing. All cell clumps should be broken up.
5. Pellet the cells by centrifugation at 14,000 rpm for 30 seconds.
6. Pour off the supernatant into a container of disinfectant solution.
7. Resuspend the cell pellet in 1 ml of 100 mM LiAc.
8. Incubate the cell suspension at 30°C for 5 minutes.
9. Pellet the cells by centrifugation at 14,000 rpm for 30 seconds.
10. Pour off the supernatant into a container of disinfectant solution.
11. To the cell pellet, add the following reagents in the listed order:
    240 μl of 50% PEG
    36 μl of 1M LiAc
    50 μl of carrier DNA (2.0 mg/ml)
    5 μl of pARC145G (0.2 μg/μl)
    5 μl of pARC1520 (0.2 μg/ml)
    20 μl of sterile dH_2O
12. Vortex the mixture to resuspend the cells and mix the reagents. All cell clumps should be broken up.
13. Incubate at 42°C for 20 minutes.
14. Pellet the cells by centrifugation at 14,000 rpm for 30 seconds.
15. Remove the supernatant using a micropipette and discard it into disinfectant solution.
16. Resuspend the cell pellet in 200 μl of sterile dH_2O by *slowly and gently* pipetting the mixture up and down.

PROTOCOL 2 PLATING THE YEAST TRANSFORMANTS

1. Obtain the four different types of SC agar plates.
2. Label each plate with your initials and the date.
3. Pipette 50 µl of transformed yeast onto each plate.
4. Using a sterile cell spreader, spread the yeast over the surface of the agar. Move the spreader back and forth while turning the plate.
5. Incubate the plates upside down in a 30°C incubator for 3 to 4 days.
6. Observe and record the number of yellow and pink colonies.

Full pigment production may require more than 4 days.

References and General Reading

Armstrong, G. A., and J. E. Hearst. 1996. Carotenoids 2: genetics and molecular biology of carotenoid pigment biosynthesis. *FASEB J.* **10:**228–237.

Ausubel, F. M., R. Brent, R. E. Kingston, D. D. Moore, J. G. Seidman, J. A. Smith, and K. Struhl (ed.). 1994. *Current Protocols in Molecular Biology.* Wiley Interscience, New York, N.Y.

Brown, A. J. P., and M. F. Tuite (ed.). 1998. *Methods in Microbiology,* volume 26. *Yeast Gene Analysis.* Academic Press Inc., New York, N.Y.

Demain, A. L. 1999. Stunning achievements of industrial microbiology. *ASM News* **65:**311–316.

Glick, B. R., and J. J. Pasternak. 1998. *Molecular Biotechnology: Principles and Applications of Recombinant DNA,* 2nd ed. ASM Press, Washington, D.C.

Hundle, B., M. Alberti, V. Nievelstein, P. Beyer, H. Kleinig, G. A. Armstrong, D. H. Burke, and J. E. Hearst. 1994. Functional assignment of *Erwinia herbicola* Eho10 carotenoid genes expressed in *Escherichia coli. Mol. Gen. Genet.* **245:**406–416.

Hundle, B. S., P. Beyer, H. Kleinig, G. Englert, and J. E. Hearst. 1991. Carotenoids of *Erwinia herbicola* and an *Escherichia coli* HB101 strain carrying the *Erwinia herbicola* carotenoid gene cluster. *Photochem. Photobiol.* **54:**89–93.

Hundle, B. S., D. A. O'Brien, P. Beyer, H. Kleinig, and J. E. Hearst. 1993. *In vitro* expression and activity of lycopene cyclase and beta-carotene hydroxylase from *Erwinia herbicola. FEBS Lett.* **315:**329–334.

Misawa, N., and H. Shimada. 1997. Metabolic engineering for the production of carotenoids in non-carotenogenic bacteria and yeasts. *J. Biotechnol.* **59:**169–181.

Miura, Y., K. Kondo, T. Saito, H. Shimada, P. D. Fraser, and N. Misawa. 1998. Production of the carotenoids lycopene, beta-carotene, and astaxanthin in the food yeast *Candida utilis. Appl. Environ. Microbiol.* **64:**1226–1229.

Miura, Y., K. Kondo, H. Shimada, T. Saito, K. Nakamura, and N. Misawa. 1998. Production of lycopene by the food yeast, *Candida utilis,* that does not naturally synthesize carotenoid. *Biotechnol. Bioeng.* **58:**306–308.

Paetkau, D. W., J. A. Riese, W. S. MacMorran, R. A. Woods, and R. D. Gietz. 1994. Interaction of the yeast RAD7 and SIR3 proteins: implications for DNA repair and chromatin structure. *Genes Dev.* **8:**2035–2045.

Perry, K. L., T. A. Simonitch, K. J. Harrison-Lavoie, and S. T. Liu. 1986. Cloning and regulation of *Erwinia herbicola* pigment genes. *J. Bacteriol.* **168:**607–612.

Tuveson, R. W., and G. Sandmann. 1993. Protection by cloned carotenoid genes expressed in *Escherichia coli* against phototoxic molecules activated by near-ultraviolet light. *Methods Enzymol.* **214:**323–330.

Interesting Internet Sites

http://dcb-carot.unibe.ch
http://www.lycopene.org
http://www.urmc.rochester.edu/smd/biochem/yeast/10.html

plant cell by a mechanism thought to resemble bacterial conjugation. The right and left border sequences are the specific cleavage sites for release of these sequences and the T-DNA between them. In addition, one or both border sequences are required for T-DNA integration into the plant genome. In the next step of tumor production, the T-DNA genes for auxin and cytokinin (plant growth hormones) are activated, generating the rapid, uncontrolled growth characteristic of a crown gall tumor.

Other Ti plasmid genes include those coding for the synthesis of opines, unique derivatives of amino acids that serve as nutrients for *A. tumefaciens* (Fig. 21.2). For example, nopaline, one of the opines, is the result of a condensation reaction between arginine and α-ketoglutaraldehyde. Nopaline synthase, the enzyme involved in the reaction, has been used as a reporter gene in plant transformation experiments. Plants produce and secrete opines only in response to infection and lack the genes necessary to catabolize them. *A. tumefaciens*, on the other hand, can obtain energy from opine breakdown, since its Ti plasmid also possesses opine catabolism genes. Infected plant cells spend a considerable portion of their energy to support *A. tumefaciens* metabolism at their own expense.

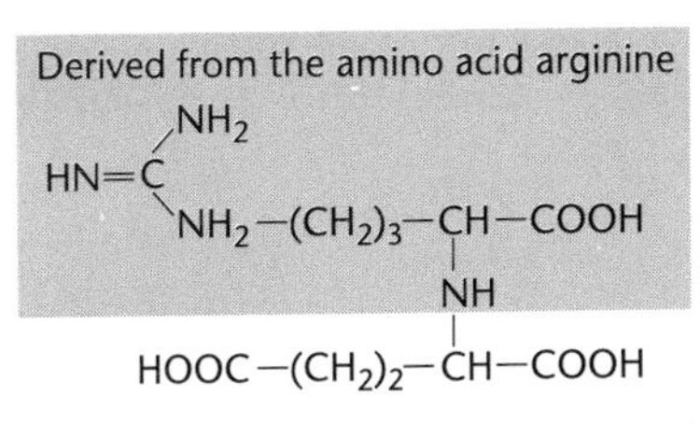

Figure 21.2 Chemical structure of nopaline, an opine.

Numerous genes of interest, such as those for insect toxins, have been inserted into the Ti plasmid and transformed into plant cells. However, Ti plasmids as they exist in nature are not good cloning vectors. To make the plasmid more useful, the auxin, cytokinin, and opine synthesis genes are deleted to eliminate pathogenesis. Smaller plasmids are more desirable for recombinant DNA experiments, so nonessential regions of the plasmid are also removed. An *E. coli* origin of replication (ori) is inserted, which facilitates cloning, manipulation, and maintenance of genes of interest in *E. coli* host cells. Plant selectable and bacterial selectable marker genes, a multiple cloning site, and the right border of the T-DNA are also inserted. The vector thus constructed is lacking the *vir* genes necessary for transfer of DNA into the plant cell. Most Ti plasmid vector systems actually use two plasmids. The cloning vector contains the gene of interest, the plant and bacterial selectable markers, either the *E. coli* ori or both the *E. coli* and *A. tumefaciens ori* sequences, and either the right or both left and right border sequences. The second plasmid resides within the recipient *A. tumefaciens* cell. It always contains the *vir* genes and may contain essential components not present in the cloning vector, but it is disarmed, lacking part or all of the T-DNA. The cloning vector is transformed into an *A. tumefaciens* strain possessing the disarmed resident plasmid, and the transformed cells are used to infect the plant tissue. The *vir* genes of the disarmed plasmid direct insertion of the cloning vector T-DNA into the plant cell.

In this exercise, we are using *A. tumefaciens* ATCC 33970. We will create "wounds" by cutting leaf tissue disks with a sterile cork borer. The resulting damaged plant tissue will be incubated with *A. tumefaciens*. Successful transformation will be assessed by paper chromatography of leaf lysate; the chromatographs will be sprayed with ninhydrin to detect amino acids and amino acid derivatives. The amino acid pattern will be compared with that of a pure nopaline control. The presence of nopaline provides conclusive evidence of plant cell transformation with the Ti plasmid.

Lab Language

B. thuringiensis toxin
callus
chromatography
crown gall tumor
dicot
glyphosate
macerate
meristem cell
microprojectile bombardment
monocot
ninhydrin
nopaline
opines
Ti plasmid
totipotent

Laboratory Safety

Review the sections on physical, biological, and chemical hazards in "Laboratory Safety." Special considerations for this exercise:

hot, autoclaved solutions
use of microbes
ninhydrin reagent

Materials

Class equipment and supplies

75% ethanol
12% bleach
aluminum foil
balance
microcentrifuge
chromatography solvent
- formic acid (120 ml)
- isopropanol (700 ml)
- dH_2O (180 ml)

ninhydrin reagent in spray bottles (0.2 g of ninhydrin dissolved in 100 ml of 95% ethanol)
handheld hair dryer
purified sea sand grinding medium
sterile dH_2O
waste containers
chemical fume hood

Materials per pair of students

potato, tomato, or tobacco leaf
sterile empty petri dishes
10-mm cork borer
50-ml sterile tubes
2 plates of M-S (Murashige and Skoog) medium
1 aliquot of *A. tumefaciens* culture
ice-cold mortar and pestle
microcentrifuge tubes
10- to 100-µl micropipette and sterile tips
chromatography paper (15 by 20 cm)
pencil
metric ruler
sterile forceps
disposable exam gloves
chromatography chamber
1 aliquot of nopaline

Protocols

PROTOCOL 1 TRANSFORMATION OF PLANT CELLS

1. Obtain a small potato, tomato, or tobacco leaf.

Ethanol and bleach serve to disinfect the leaf surface.

2. Place the leaf into a 50-ml culture tube and add sufficient 75% ethanol to cover the leaf. Cap the tube.
3. Incubate at room temperature for 10 minutes.
4. Decant the ethanol into a suitable waste container. Use sterile forceps to retain the leaf in the tube, if necessary.
5. Add sufficient 12% bleach to cover the leaf.
6. Incubate for 5 minutes at room temperature.
7. Decant the bleach solution into a suitable waste container.

Rinsing removes residual ethanol and bleach that could interfere with the experiment.

8. Cover the leaf with distilled water (dH_2O) to rinse. Decant the water.
9. Repeat step 8 four more times.
10. Remove the leaf from the tube and place it into a sterile empty petri dish.
11. Using a sterile 10-mm cork borer, punch 20 disks from the leaf.

12. Obtain two M-S media plates. Label one "control" and the other "transformation."

13. Using sterile forceps, place 10 leaf disks on the control plate, equally spaced.

14. Using sterile forceps, dip 10 leaf disks into the *A. tumefaciens* culture, then place each onto the transformation plate, equally spaced.

15. Wrap each plate in aluminum foil and incubate at room temperature for 48 hours.

EXERCISE 21

M-S (Murashige and Skoog) medium is a defined medium high in sucrose that is widely used for plant biotechnology protocols.

Temperature is critical: plasmid replication is inhibited above 30°C.

PROTOCOL 2 PREPARATION OF LEAF CELL LYSATE

1. Collect the leaf disks from the transformation plate.

2. Weigh out 0.5 g of leaf tissue and place it into an ice-cold mortar.

3. Add a minute amount of sea sand, and macerate to a paste using an ice-cold pestle.

4. Transfer the paste to a 1.5-ml microcentrifuge tube, and centrifuge at 14,000 rpm for 5 minutes.

5. Label a clean 1.5-ml microcentrifuge tube "transformed 1."

6. With a 10- to 100-μl micropipette and sterile tip, carefully remove the supernatant and transfer it to the labeled microcentrifuge tube.

7. Repeat steps 2 through 4, then label a clean 1.5-ml microcentrifuge tube "transformed 2."

8. With a 10- to 100-μl micropipette and sterile tip, carefully remove the supernatant from the second sample and transfer it to the second labeled microcentrifuge tube.

9. For your negative control, repeat steps 1 through 8 using the leaf tissue collected from the control plate.

Disposable microtube pestles and microcentrifuge tubes may be suitable.

Sea sand assists in rupturing the cell walls.

PROTOCOL 3 PAPER CHROMATOGRAPHY OF LEAF LYSATES

1. **Wearing gloves,** obtain a sheet of chromatography paper. *Handle by the corners only.* Place the chromatography paper on a sheet of paper towel.

2. Using a pencil and a ruler, lightly draw a line parallel to the long edge of the paper 2.5 cm from the edge.

3. Using a pencil, make five small, equally spaced guide marks on the line, as illustrated in Fig. 21.3.

4. Label the marks as follows: N, nopaline control; T1, transformed sample 1; T2, transformed sample 2; C1, control sample 1; C2, control sample 2.

5. Obtain 1 aliquot of nopaline solution.

6. Using a clean 10- to 100-μl micropipette tip for the nopaline control and each supernatant collected in protocol 2, add 15 μl of each sample to the appropriately marked spot on the paper. Allow the spots to dry thoroughly.

Gloves and careful handling will minimize contamination from amino acids on the skin.

Add the sample in several tiny drops, allowing each to dry before adding additional solution.

Figure 21.3 Guide diagram for paper chromatography of leaf lysates.

Allow about 5 minutes for the solvent vapor to fill the jar so that the liquid and vapor phases will equilibrate before you insert your chromatogram.

7. Add sufficient chromatography solvent to the chromatography chamber to form a 2-cm-high solvent layer. Cover the chamber.

8. Staple the ends of the paper to form a nonoverlapping cylinder so that the samples are on the outside of the cylinder and at the bottom, as illustrated in Fig. 21.4.

9. Place your chromatogram in the chamber, *with the samples at the bottom.* Close the chamber.

The chromatography will take about 2 hours.

10. Allow the solvent front to move up the paper until it reaches about 2 cm from the top of the paper.

The solvent front position must be marked immediately, since it will not be visible when the paper is dry.

11. Remove the chromatogram from the chamber and immediately remove the staples and draw a pencil line to mark the position of the solvent front.

12. Allow the chromatogram to dry.

13. **Wearing gloves** and working in the chemical fume hood, spray your chromatogram with ninhydrin reagent.

Heat will reveal amino acids and amino acid derivatives as bluish spots on the paper.

14. Dry the sprayed chromatogram with a handheld hairdryer set on high heat.

PROTOCOL 4 CALCULATION OF R_f VALUES

1. With a centimeter ruler, measure the distance from the line marking the original position of the samples to the position of the solvent front, in millimeters.

The nopaline standard serves as a positive control.

2. Measure the distance the nopaline standard and each sample migrated from the origin, in millimeters.

Figure 21.4 Paper chromatography apparatus. (A) Paper chromatogram cylinder showing positions of samples and staples. (B) After equilibrium of liquid and gas phases, the chromatogram is inserted in the closed chamber. (C) A glass plate is used to cover and seal the chamber.

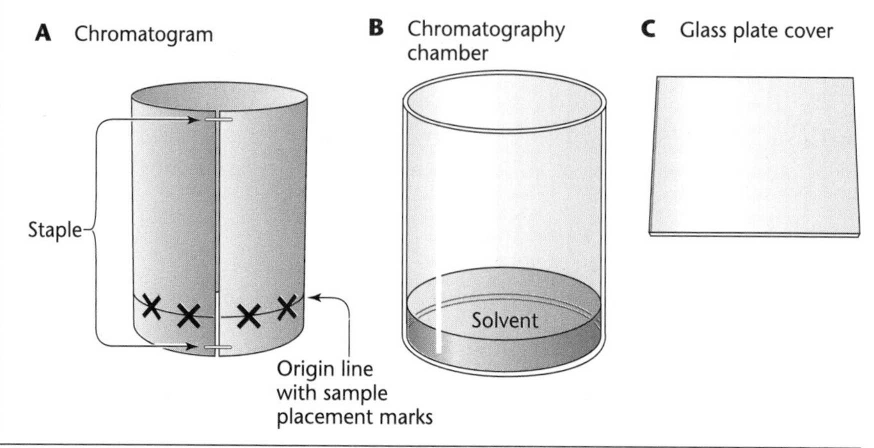

EXERCISE 21

3. Calculate the R_f value for the nopaline standard. R_f = distance nopaline migrated from origin (mm)/distance solvent front migrated from origin (mm).

R_f stands for the ratio of fronts, where the distance of the substance front is compared with the distance of the solvent front.

4. Calculate the R_f values for the amino acids and amino acid derivatives in your T1, T2, C1, and C2 lanes. R_f = distance sample migrated from origin (mm)/distance solvent front migrated from origin (mm).

Several different spots may be visible in the lanes.

5. Identify the position of nopaline in the T1 and T2 lanes.

References and General Reading

Bilang, R., and M. Schrott. 1995. Genetic manipulation of plant cells. *In* R. A. Meyers (ed.), *Molecular Biology and Biotechnology: a Comprehensive Desk Reference.* VCH Publishers, New York, N.Y.

Clark, D. P., and L. D. Russell. 1997. *Molecular Biology Made Simple and Fun.* Cache River Press, Vienna, Ill.

DellaPenna, D. 1999. Nutritional genomics: manipulating plant micronutrients to improve human health. *Science* **285:**375–379.

Glick, B. R., and J. J. Pasternak. 1998. *Molecular Biotechnology: Principles and Applications of Recombinant DNA,* 2nd ed. ASM Press, Washington, D.C.

Lebowitz, R. J. 1995. *Plant Biotechnology: a Laboratory Manual.* William C. Brown, Dubuque, Iowa.

Lodish, H., D. Baltimore, A. Berk, S. L. Zipursky, P. Matsudaira, and J. Darnell. 1995. *Molecular Cell Biology,* 3rd ed. W. H. Freeman & Co., New York, N.Y.

Mazur, B., E. Krebbers, and S. Tingey. 1999. Gene discovery and product development for grain quality traits. *Science* **285:**372–375.

Oard, J. 1995. Physical methods for plant cell transformation. *In* R. A. Meyers (ed.), *Molecular Biology and Biotechnology: a Comprehensive Desk Reference.* VCH Publishers, New York, N.Y.

Ronald, P. C. 1997. Making rice disease-resistant. *Sci. Am.* **276:**100–105.

Vidaver, A. K. 1999. Plant microbiology: century of discovery, with golden years ahead. *ASM News* **65:**358–363.

Discussion

1. Discuss factors that make it difficult to transform plant cells.
2. Discuss the *A. tumefaciens* Ti plasmid and its role in plant pathogenesis.
3. Describe microprojectile bombardment.
4. Why must you use leaf disks rather than whole leaves for transformation?
5. What is the significance of nopaline and other opines?
6. Why is nopaline a marker for successful transformation?
7. What is the R_f value of nopaline?
8. How many different spots did you find on your chromatogram?
9. How can you identify nopaline in your samples?
10. Compare your chromatograms and R_f values with those of your classmates. How do they differ? Explain.

SECTION VII

Techniques for Studying Evolution and Systematics

EXERCISE 22

Isolation of Genomic DNA

Key Concepts

1. Cell and/or tissue organization varies according to each species' phylogenetic position. Different protocols for DNA and protein isolation are required for various taxa.
2. Before DNA can be isolated from a multicellular organism, its tissue must be dissociated into individual cells.

Goals

By the end of this exercise, you will be able to

- isolate DNA from bacterial, *Euglena*, yeast, plant, chicken, and human cells
- identify properties of cell types that must be considered when isolating DNA

Introduction

A thorough understanding of cell biology is fundamental to molecular biologists. One must comprehend the diversity of living organisms to find clues to potential applications in biotechnology. For example, extraction of DNA from a plant cell requires protocols somewhat different from those needed to extract DNA from a bacterial cell. Bacteria are unicellular and prokaryotic; their cells are small and lack a nucleus and complex membranous organelles. Plants, on the other hand, are eukaryotic and multicellular; plant cells have a thick, hard cellulose wall, and a double membrane surrounds the nucleus. The techniques to extract DNA from both cell types are similar, but not identical. Manipulation of these two organisms varies, dictated by constraints of cell and tissue structure.

Every organism is composed of one of two structurally different types of cells: prokaryotic or eukaryotic. Prokaryotic cells evolved about 3.5 billion years ago; eukaryotic cells followed about 2 billion years later. According to the endosymbiotic theory proposed and expanded by Lynn Margulis, eukaryotic cells arose through close relationships between several prokaryotic cells, in which one or more prokaryotic cells came to be included within the cytoplasm of a larger prokaryotic cell. Eventually, these endosymbionts gave rise to mitochondria and chloroplasts. Nuclei may have arisen via invagination of the plasma membrane. *Streptococcus pyogenes,* the bacterium that causes strep throat, is an example of a prokaryote. *Saccharomyces cerevisiae,* or brewer's yeast, the organism that makes bread rise and ferments barley to beer, is an example of a unicellular eukaryote. Dogs and tulips are examples of multicellular eukaryotes. The sequences of mitochondrial DNA and chloroplast DNA offer evidence for the endosymbiotic theory of the origin of eukaryotes.

The Classification of Living Organisms

As in the 18th century, when Carolus Linnaeus first relegated all living organisms to either the plant kingdom or the animal kingdom, taxonomy still is a work in progress. Scientists have built present taxonomic schemes to reflect the very real similarities and differences in cell structure and function found among living organisms.

In 1969, Robert Whittaker presented a five-kingdom system of classification, later elaborated by Lynn Margulis and Karlene Schwartz. The five-kingdom system comprises a single prokaryotic kingdom, kingdom *Monera,* or *Procaryotae,* and four eukaryotic kingdoms. Within these eukaryotic kingdoms, all unicellular and simple multicellular organisms are grouped in the kingdom *Protista,* or *Protoctista.* The other three kingdoms are multicellular, and their differences are based on their types of cells, modes of acquiring food (chemical bond energy), development, and level of cell specialization. The multicellular kingdoms are kingdom *Fungi,* kingdom *Plantae,* and kingdom *Animalia.*

However, an alternative system, initially proposed by Carl Woese, categorizes living organisms into one of only three kingdoms or domains. These are the *Archaea,* the *Bacteria,* and the *Eucarya.* Recent evidence from ribosomal RNA (rRNA) sequence studies supports the very close evolutionary relationships among all the eukaryotic lineages. However, rRNA data indicate that the archaebacteria (*Monera*), or *Archaea,* are fundamentally different from the eubacteria (*Monera*), or *Bacteria,* and these two types of organisms should be grouped into separate kingdoms or domains. Data also indicate that eukaryotic organisms may be more closely related to *Archaea* than to *Bacteria.*

Because of its widespread acceptance and easily utilizable structure, we will employ the five-kingdom system as the framework for our discussion and genomic DNA extraction and purification protocols.

Kingdom Monera

The kingdom *Monera* includes all prokaryotic organisms. These are very small (1 to 10 μm) and tend to live in colonies. Colonies are aggregates of organisms living together, but each cell of the colony is an independent organism with its own reproductive and metabolic functions.

Monerans can be divided into two main groups: the eubacteria, or true bacteria, and the archaebacteria, or ancient bacteria. Eubacteria include a variety of organisms: the bacteria associated with humans, such as *Escherichia coli* and *S. pyogenes*; photosynthesizers, such as the oxygen-releasing cyanobacteria; and the several species responsible for the nitrogen cycle in ecosystems. The archaebacteria enjoy an extremophilic lifestyle. They are found in a variety of extreme environmental habitats, such as very high salinity, very high temperature, extreme acidity or alkalinity, and so forth.

The basic metabolic pathways, found throughout the five kingdoms, for heterotrophic synthesis of ATP from respiration of organic substrates evolved in the monerans. Photosynthesis also arose in the *Monera.* However, some monerans metabolize in a variety of ways not found in the other kingdoms. For example, there are bacteria that obtain energy by oxidizing iron, sulfur, or ammonia.

Kingdom Protista *or* Protoctista

Protistan systematics is perhaps the most fluid and hotly debated of the four eukaryotic kingdoms. The members of the kingdom *Protista,* or *Protoctista,* include all the eukaryotic single-cell organisms as well as simple multicellular organisms. The abundance of organisms with these characteristics has led to differentiating them based on their cell structure and metabolism type. The three main groups of protistans are the algae, the protozoa, and the slime molds.

Algae are plantlike, possessing a cell wall and chloroplasts. They are photosynthetic autotrophs. *Euglena, Chlorella,* and *Spirogyra* spp. are typical examples of algae. You may also be familiar with other organisms of this group, such as the green filaments commonly seen in creeks and ponds and the brown and red seaweeds found in saltwater habitats.

Protozoa have an animallike cell that lacks cell walls and plastids. These organisms are heterotrophs: typically, they ingest their food. *Amoeba* and *Paramecium* spp. are representative examples of these protists.

Slime molds are the least common protists. A funguslike cell characterizes them. They are heterotrophs, but unlike the protozoa, slime molds absorb their food as dissolved organic substrates. The slimy yellowish or brownish substances often found in decomposing wood generally are the species of *Physarum*, a well-known slime mold.

Kingdom Fungi

Most of the organisms within the kingdom *Fungi* are multicellular, but the fungi also include unicellular yeasts. The fungal cell is characterized by the presence of a cell wall composed of the strong structural polysaccharide chitin, bearing no resemblance to the cell wall of plants. These organisms are heterotrophs: some decompose organic matter, some simply metabolize dissolved organic substances found in their environment, and some are parasites.

Representatives of this kingdom are molds, mildews, mushrooms, and yeasts. The common brewer's yeast, *S. cerevisiae,* is a typical unicellular fungus. Before Whittaker's five-kingdom classification system was described, the presence of a cell wall and sedentary habits led scientists to classify mildews and mushrooms as plants. However, it is clear that these organisms have nothing in common with plants, and they have been placed in a separate kingdom.

Kingdom Plantae

Plants are multicellular photoautotrophic organisms. They develop from embryos protected within maternal tissue. The cells of plants are organized into specialized tissues that allow a division of labor between the different portions of an organ and between various organs. Examples of these tissues are photosynthetic tissue (photosynthetic parenchyma), transport tissues (xylem and phloem), and storage tissue (storage parenchyma). An excellent example of this complexity is observed in the leaf of elodea (*Anacharis*): a simple layer of transparent cells, the epidermis, covers the leaf. Inside the leaf are mesophyll cells specialized for photosynthesis and cells of xylem and phloem specialized for the transport of water, minerals, and organic substances. Plant organs include roots, stems, and leaves.

Kingdom Animalia

Animals are multicellular, eukaryotic, and heterotrophic organisms. Typically, they obtain food by ingestion. Animals include a blastula in their early development. Like plants, animal cells form specialized tissues of great complexity, and tissues of various types are arranged into organs. For example, a single layer of endothelial cells lines arteries; smooth muscle cells are found within the artery wall; and connective tissue forms the outer coating. Arteries, veins, capillaries, and the heart are organs of the circulatory system.

Genomic DNA Extraction

Isolation and purification of DNA are essential procedures in molecular biology. Extraction and purification of chromosomal or genomic DNA from prokaryotic and eukaryotic organisms have facilitated the study of complex genomes, particularly by allowing the construction of genomic DNA libraries for many species. Extraction and purification techniques provide DNA suitable for a variety of applications, including amplifications by polymerase chain reaction, digestion with restriction endonucleases, and membrane hybridizations, such as Southern blots.

Chemically, DNA is identical from one organism to another. It is the sequence of base pairs that distinguishes DNA from various species. Without the property of identity, it would be impossible for recombinant DNA technology to be carried out; bacteria would be unable to replicate foreign DNA engineered into plasmids if it did not have the same basic structure as native DNA. The composition of an organism, whether it is unicellular or multicellular, and the exact makeup of tissues and individual cells affect the protocols used to isolate DNA. Also, the use for which the DNA is required may dictate the method employed in its isolation.

The basic structure of DNA is discussed and explored in a number of exercises in this book. In exercise 10, DNA is isolated from human cheek cells by a quick method. Plasmid DNA is isolated from bacteria in exercise 8, and large intact pieces of human genomic DNA are isolated for a Southern blot in exercise 16. Each isolation method was chosen to suit the specific application and the nature of the DNA.

This exercise isolates DNA from organisms representative of the five kingdoms: *Animalia* (humans), *Fungi* (yeasts), *Plantae* (plants), *Protista* (euglena), and *Monera* (bacteria). The presence or absence of a nucleus as well as the makeup of the cell plays a role in determining the DNA isolation scheme. Although the basic extraction and purification protocols are similar for all of these organisms, both plants and yeast require additional steps because of their cell wall compositions.

The genomic DNA purification technique that we will use in this exercise is based on a four-step process (Fig. 22.1).

- Cells and nuclei, if present, must be lysed to release genomic DNA. Lysis solutions commonly include the detergent sodium dodecyl sulfate (SDS) and sodium hydroxide.
- Ribonuclease (RNase) is added to the cell lysate to remove RNA from the sample.
- The cellular proteins are then removed by addition of ammonium or potassium acetate, which precipitates the proteins but leaves the high-molecular-weight genomic DNA in solution.
- Finally, the genomic DNA is concentrated and desalted by isopropanol precipitation, followed by an ethanol wash.

Lab Language

endosymbiotic theory
systematics
taxa

Laboratory Safety

Review the sections on physical, biological, and chemical hazards in "Laboratory Safety." Special considerations for this exercise:

working with microbes
warm solutions
boiling agarose
gel electrophoresis
staining with ethidium bromide
ultraviolet transilluminator

Laboratory Sequence

Day 1: DNA isolation from bacteria, *Euglena gracilis,* plants or algae, and cheek cells or chicken erythrocytes
Begin DNA isolation from yeasts.
Prepare gels for electrophoresis during incubations.

Day 2: Gel electrophoresis
Quantification of DNA

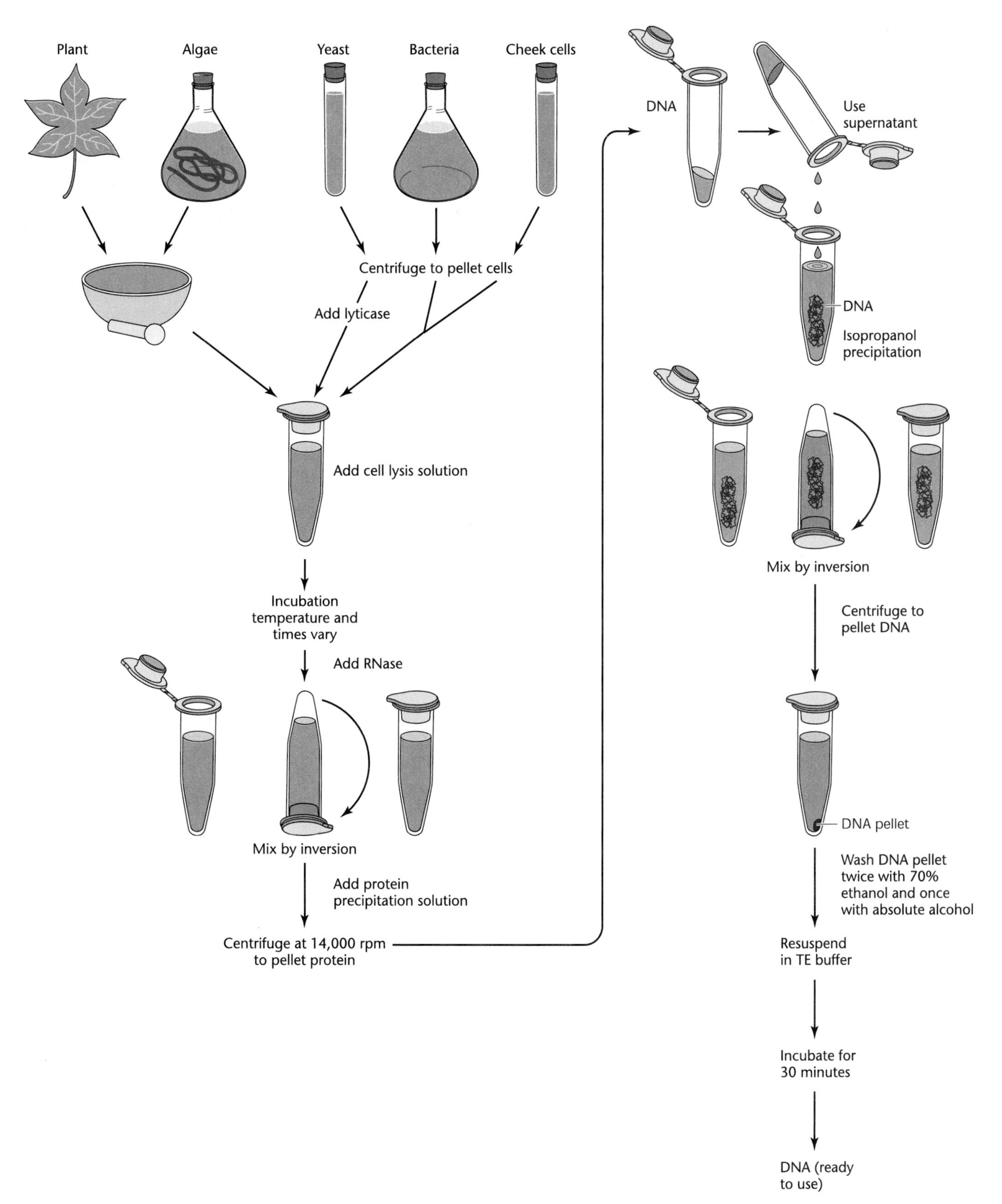

Figure 22.1 Summary of genomic DNA extraction from various organisms.

Materials

Class equipment and supplies
overnight *E. coli* broth culture
overnight *S. cerevisiae* culture
Euglena culture
chicken erythrocytes
20-mg/ml lyticase
50 mM EDTA, pH 8.0
cell lysis solution (Tris-EDTA-SDS) from Puregene (Gentra Systems)
protein precipitation solution (ammonium acetate) from Puregene (Gentra Systems)
1× Tris-EDTA hydration solution, or from Puregene (Gentra Systems)
100% isopropanol
70% ethanol
clinical centrifuge
microcentrifuge
37°C water bath
65°C water bath
80°C heat block
vortex
microcentrifuge
ultraviolet transilluminator
ethidium bromide staining solution
Polaroid camera and film

Materials per pair of students
2 plastic medicine cups
2 paper cups
2 15-ml conical centrifuge tubes
microcentrifuge tubes
0.5- to 10-µl micropipette and sterile tips
10- to 100-µl micropipette and sterile tips
100- to 1,000-µl micropipette and sterile tips
chilled mortar and pestle or toothpick
1 aliquot of cell lysis solution
1 aliquot of RNase A
1 aliquot of protein precipitation solution
1 aliquot of 100% isopropanol
1 aliquot of 70% ethanol
1 aliquot of hydration solution
gel electrophoresis chamber, gel casting tray, comb
power supply (two pairs of students will share this)
1 aliquot of uncut lambda DNA marker

Students will need to prepare
0.9% NaCl solution (exercise 3)
1× Tris-borate-EDTA buffer for electrophoresis

Protocols

PROTOCOL 1 GENOMIC DNA EXTRACTION FROM GRAM-NEGATIVE BACTERIA

1. Obtain an overnight broth culture of *E. coli.*
2. Label a microcentrifuge tube. Transfer approximately 1.0 ml of the culture into the tube, using a Pasteur pipette.
3. Pulse the tube in the microcentrifuge for 5 to 10 seconds to pellet the bacteria.
4. Carefully remove most of the supernatant by pouring it into a container of disinfectant.
5. Add 600 µl of cell lysis solution and gently pipette up and down to resuspend cells.

SDS, a detergent in the lysis solution, breaks open the cell membranes.

6. Incubate at 80°C for 5 minutes.
7. Cool the sample at room temperature and add 3 µl of RNase A solution to the cell lysate.

The sample must be cool; heat will denature the RNase enzyme.

8. Mix by inverting the tubes 25 times, and incubate for 30 minutes at 37°C.

9. Cool the sample at room temperature. Add 200 µl of protein precipitation solution and vortex vigorously for 20 seconds.

Ammonium acetate causes the proteins to precipitate, leaving the DNA in solution.

10. Microcentrifuge the sample at 14,000 rpm for 3 minutes to pellet the protein debris.

11. Pour the supernatant into a clean 1.5-ml tube, leaving the protein debris behind.

12. Add 600 µl of 100% isopropanol and mix *gently* by inverting the tube at least 50 times.

Vigorous mixing will shear DNA into small fragments.

13. Centrifuge at 14,000 rpm for 1 minute to pellet the DNA.

Orient the hinge on the microcentrifuge tube outward to assist in locating the tiny DNA pellet.

14. Pour off supernatant and drain the liquid on an absorbent towel.

The pellet may appear off-white to yellow.

15. Add 600 µl of 70% ethanol and invert the tube several times to wash the pelleted DNA.

A 70% ethanol wash removes unwanted salts from the DNA.

16. Centrifuge at 14,000 rpm for 1 minute.

17. Pour off the supernatant, *slowly and carefully*, watching the pellet so it does not pour out.

18. Invert the tube and drain all of the ethanol. Air dry the DNA pellet for at least 15 minutes.

Any residual ethanol will interfere with subsequent analysis.

19. Add 100 µl of hydration solution to the pellet and place the tube in a water bath at 65°C for 1 hour or incubate overnight at room temperature.

PROTOCOL 2 GENOMIC DNA EXTRACTION FROM *EUGLENA GRACILIS*

1. Obtain a pure broth culture of *E. gracilis*.

2. Label a microcentrifuge tube. Transfer approximately 1.5 ml of the culture into the tube, using a Pasteur pipette.

3. Microcentrifuge the tube at 10,000 rpm for 10 seconds to pellet the *E. gracilis*.

4. Carefully remove most of the supernatant by pouring it into a container of disinfectant.

5. Add 600 µl of cell lysis solution and gently pipette up and down to resuspend cells.

SDS, a detergent in the lysis solution, breaks open the cell membranes.

6. Incubate at 65°C for 1 hour.

7. Cool the sample at room temperature and add 3 µl of RNase A solution to the cell lysate.

The sample must be cool; heat will denature the RNase enzyme.

8. Mix by inverting the tubes 25 times, and incubate for 15 minutes at 37°C.

9. Cool the sample at room temperature. Add 200 µl of protein precipitation solution and vortex vigorously for 20 seconds.

Ammonium acetate causes the proteins to precipitate, leaving the DNA in solution.

10. Microcentrifuge the sample at 14,000 rpm for 3 minutes to pellet the protein debris.

The pellet will appear greenish because of chlorophyll.

11. Pour the supernatant into a clean 1.5-ml tube, leaving the protein debris behind.

Vigorous mixing will shear DNA into small fragments.

12. Add 600 µl of 100% isopropanol and mix *gently* by inverting the tube at least 50 times.

Orient the hinge on the microcentrifuge tube outward to assist in locating the tiny DNA pellet.

13. Centrifuge at 14,000 rpm for 1 minute to pellet the DNA.

The pellet may appear off-white to green.

14. Pour off the supernatant and drain the liquid on an absorbent towel.

A 70% ethanol wash removes unwanted salts from the DNA.

15. Add 600 µl of 70% ethanol and invert the tube several times to wash the pelleted DNA.

16. Centrifuge at 14,000 rpm for 1 minute.

17. Pour off the supernatant, *slowly and carefully,* watching the pellet so it does not pour out.

Any residual ethanol will interfere with subsequent analysis.

18. Invert the tube and drain all of the ethanol. Air dry the DNA pellet for at least 15 minutes.

19. Add 100 µl of hydration solution to the pellet, and place the tube in a water bath at 65°C for 1 hour or incubate overnight at room temperature.

PROTOCOL 3 GENOMIC DNA EXTRACTION FROM PLANT TISSUES AND/OR *SPIROGYRA* SPECIES

For plant tissue use:
- *10 to 20 mg of dry tissue*
- *20 to 60 mg of frozen tissue*
- *20 to 60 mg of fresh tissue*

For Spirogyra, use:
- *50 mg of the algae thread*

1. Collect fresh leaf tissue or obtain a *Spirogyra* culture. Tender tissues are best for DNA extraction. Use it immediately, freeze it in liquid nitrogen, or dry it at room temperature.

2. Grind the fresh tissue using a chilled mortar and pestle. If the plant tissue is fresh and tender, homogenize it in a microcentrifuge tube using a toothpick or small pestle.

3. If necessary, place the plant tissue in a 1.5-ml microcentrifuge tube.

SDS, a detergent in the lysis solution, breaks open the cell membranes.

4. Add 600 µl of cell lysis solution and gently pipette up and down to resuspend cells.

5. Incubate at 65°C for 1 hour.

The sample must be cool; heat will denature the RNase enzyme.

6. Cool the sample at room temperature and add 3 µl of RNase A solution to the cell lysate.

7. Mix by inverting the tubes 25 times, and incubate for 15 minutes at 37°C.

Ammonium acetate causes the proteins to precipitate, leaving the DNA in solution.

8. Cool the sample at room temperature. Add 200 µl of protein precipitation solution and vortex vigorously for 20 seconds.

The pelleted protein debris will appear greenish.

9. Microcentrifuge the sample at 14,000 rpm for 3 minutes to pellet the protein debris.

10. Pour the supernatant into a clean 1.5-ml tube, leaving the protein debris behind.

11. Add 600 µl of 100% isopropanol and mix *gently* by inverting the tube at least 50 times.

Vigorous mixing will shear DNA into small fragments.

12. Centrifuge at 14,000 rpm for 1 minute to pellet the DNA.

Orient the hinge on the microcentrifuge tube outward to assist in locating the tiny DNA pellet.

13. Pour off the supernatant and drain the liquid on an absorbent towel.

The pellet may appear off-white to green.

14. Add 600 µl of 70% ethanol and invert the tube several times to wash the pelleted DNA.

A 70% ethanol wash removes unwanted salts from the DNA.

15. Centrifuge at 14,000 rpm for 1 minute.

16. Pour off the supernatant, *slowly and carefully*, watching the pellet so it does not pour out.

17. Invert the tube and drain all of the ethanol. Air dry the DNA pellet for at least 15 minutes.

Any residual ethanol will interfere with subsequent analysis.

18. Add 100 µl of hydration solution to the pellet and place the tube in a water bath at 65°C for 1 hour or incubate overnight at room temperature.

PROTOCOL 4 GENOMIC DNA EXTRACTION FROM YEAST CELLS (*S. CEREVISIAE*)

1. Obtain an overnight broth culture of *S. cerevisiae.*
2. Label a microcentrifuge tube. Transfer approximately 1.5 ml of the culture into the tube, using a Pasteur pipette.
3. Pulse the tube in the microcentrifuge for 5 seconds to pellet the yeast.
4. Carefully remove most of the supernatant by pouring it into a container of disinfectant.
5. Add 300 µl of 50 mM EDTA, pH 8.0, and gently pipette up and down to resuspend cells.
6. Add 7.5 µl of 20-mg/ml lyticase.

Lyticase digests chitin.

7. Mix by inverting the tube 10 times.
8. Incubate overnight at 37°C.
9. Microcentrifuge at 10,000 rpm for 5 seconds to pellet the yeast cells.
10. Carefully remove the supernatant by pouring it into a container of disinfectant.
11. Add 300 µl of cell lysis solution and pipette up and down to lyse the cells.
12. Incubate at room temperature for 5 minutes.
13. Add 3 µl of RNase A solution to the cell lysate.
14. Mix by inverting the tubes 25 times, and incubate for 15 minutes at 37°C.
15. Cool the sample at room temperature. Add 100 µl of protein precipitation solution and vortex vigorously for 20 seconds.

Ammonium acetate causes the proteins to precipitate, leaving the DNA in solution.

16. Microcentrifuge the sample at 14,000 rpm for 3 minutes to pellet the protein debris.

17. Pour the supernatant into a clean 1.5-ml tube, leaving the protein debris behind.

Vigorous mixing will shear DNA into small fragments.

18. Add 600 μl of 100% isopropanol and mix *gently* by inverting the tube at least 50 times.

Orient the hinge on the microcentrifuge tube outward to assist in locating the tiny DNA pellet.

19. Centrifuge at 14,000 rpm for 1 minute to pellet the DNA.

The pellet may appear off-white to yellow.

20. Pour off the supernatant and drain the liquid on an absorbent towel.

A 70% ethanol wash removes unwanted salts from the DNA.

21. Add 600 μl of 70% ethanol and invert the tube several times to wash the pelleted DNA.

22. Centrifuge at 14,000 rpm for 1 minute.

23. Pour off the supernatant, *slowly and carefully,* watching the pellet so it does not pour out.

Any residual ethanol will interfere with subsequent analysis.

24. Invert the tube and drain all of the ethanol. Air dry the DNA pellet for at least 15 minutes.

25. Add 100 μl of hydration solution to the pellet and place the tube in a water bath at 65°C for 1 hour or incubate overnight at room temperature.

PROTOCOL 5 OBTAINING CHEEK CELLS

An isotonic solution, 0.9% NaCl, is used to maintain cell membrane integrity.

1. Fill the plastic medicine cup with about 10 ml of sterile salt solution (0.9% NaCl). Rinse your mouth with the salt solution for at least 60 seconds. Expel the salt solution into a paper cup.

Cheek cells are constantly sloughed from the epithelium of the mouth.

2. Carefully pour the salt solution, now containing your cheek cells, into a 15-ml tube. Close the tube and centrifuge it for 10 minutes. This will force your cells to the bottom of the tube. *Keep your cup and label your tube.*

The supernatant is the liquid portion obtained after centrifugation.

The pellet is the solid part at the bottom of the tube.

3. Pour the liquid back into the paper cup. Be careful not to disturb the cells at the bottom of the tube. Remove most of the supernatant, leaving 20 to 40 μl of liquid on the cell pellet.

4. Vortex the cells and supernatant to resuspend the cell pellet.

PROTOCOL 6 GENOMIC DNA EXTRACTION FROM HUMAN CHEEK CELLS AND/OR FROM CHICKEN ERYTHROCYTES

Begin with the cheek cell pellet from protocol 5 and/or 10 μl of chicken blood.

SDS, a detergent in the lysis solution, breaks open the cell membranes.

1. Add 600 μl of cell lysis solution and gently pipette up and down to lyse the cells. Transfer all of the liquid and cells to a clean microcentrifuge tube. Label your tube.

2. Incubate the tube at 37°C until the solution is homogeneous and no cell clumps are visible. This step may be omitted if the solution resuspends without cell clumps.

RNase is an enzyme that destroys RNA.

3. Add 3 μl of RNase A solution to the cell lysate and incubate at 37°C for 30 minutes.

4. Cool the sample to room temperature.

5. Add 200 µl of protein precipitation solution and vortex vigorously for 20 seconds.

Ammonium acetate causes the proteins to precipitate, leaving the DNA in solution.

6. Microcentrifuge the sample at 14,000 rpm for 3 minutes to pellet the protein debris.

7. Pour the supernatant into a clean microcentrifuge tube, leaving the protein debris behind.

8. Add 600 µl of 100% isopropanol and mix by *gently* inverting the tube 50 times. White threads of DNA should form a visible clump.

Isopropanol causes the DNA to precipitate out of solution.

Vigorous mixing will shear DNA into smaller fragments.

9. Microcentrifuge the tube at 14,000 rpm for 1 minute. The DNA will be visible as a small white pellet.

Orient the hinge on the microcentrifuge tube outward to assist in locating the tiny DNA pellet.

10. Pour off the supernatant and drain the liquid on an absorbent towel.

11. Add 600 µl of 70% ethanol and invert the tube several times to wash the DNA.

A 70% ethanol wash removes unwanted salts from the DNA.

12. Microcentrifuge the tube at 14,000 rpm for 1 minute.

13. Pour off the ethanol, *slowly and carefully,* watching the pellet so it does not pour out.

14. Drain the liquid using an absorbent towel.

15. Allow the tube to air dry for at least 15 minutes.

Any residual ethanol will interfere with subsequent analysis.

16. Add 60 µl of DNA hydration solution and incubate at 65°C for 1 hour or overnight at room temperature to resuspend the DNA. Tap the tube occasionally to help disperse the DNA.

Do not pipette the DNA to mix, as this will cause shearing.

PROTOCOL 7 QUANTITATION OF GENOMIC DNA

DNA may be quantified by spectrophotometry or gel electrophoresis.

See exercise 12, "Quantitation of DNA by Gel Electrophoresis."

- Use a 0.8% agarose gel.
- Use 0.5-µg uncut, genomic lambda DNA as a reference standard.

Determine the approximate quantity of genomic DNA isolated by comparison of band intensity with the genomic lambda DNA.

References and General Reading

Buffone, G. J. 1985. Isolation of DNA from biological specimens without extraction with phenol. *Clin. Chem.* **31:**164–165.

Campbell, N. A., J. B. Reese, and L. G. Mitchell. 1999. *Biology,* 5th ed. Benjamin/Cummings, Menlo Park, Calif.

Gentra Systems, Inc. 1992. Puregene DNA isolation kit. Gentra Systems, Inc., Minneapolis, Minn.

Madigan, M. T., J. M. Martinko, and J. Parker. 1997. *Brock Biology of Microorganisms,* 8th ed. Prentice-Hall, Inc., Upper Saddle River, N.J.

Margulis, L. 1992. *Diversity of Life—The Five Kingdoms.* Enslow, Hillside, N.J.

Margulis, L., and K. V. Schwartz. 1988. *Five Kingdoms: an Illustrated Guide to the Phyla of Life on Earth,* 2nd ed. W. H. Freeman & Co., New York, N.Y.

Pace, N. R. 1997. A molecular view of microbial diversity and the biosphere. *Science* **276:**734–740.

Promega Corporation. 1996. *Protocols and Applications Guide,* 3rd ed. Promega Corp., Madison, Wis.

Discussion

1. Using your knowledge of various organisms' cell structure, hypothesize the reason for specific conditions used in the cell lysis step.
2. You have just identified and isolated a new organism. Design a protocol for isolating its genomic DNA. Explain your rationale.
3. How would you optimize the protocol for genomic DNA extraction from a new organism?

Examine the photo of your gel showing the isolated DNA from protocols 1 to 6.

4. Which organism resulted in the least amount of DNA? Explain.
5. Which organism yielded the most DNA? Explain.

EXERCISE 23

Comparison of DNAs from the Five Kingdoms

Key Concepts

1. Phylogenetic relationships can be determined by comparing key macromolecules.
2. Ribosomal gene sequences from different species can be used to determine evolutionary relatedness.
3. Polymerase chain reaction (PCR) can be used to amplify DNAs from different organisms that contain similar or identical sequences.

Goals

By the end of this exercise, you will be able to

- explain how PCR can amplify similar sequences from a variety of organisms
- analyze PCR products from different species
- demonstrate how small-subunit ribosomal RNA (rRNA) can be used to indicate evolutionary relatedness

Introduction

Certain cellular macromolecules, such as proteins and nucleic acids, have been used to establish molecular genealogy. Molecules, or sequences within molecules, that are widely conserved provide information about evolutionary relationships. For example, scientists have known for many years that the protein cytochrome *c* is found in all aerobic organisms, even those as diverse as *Escherichia coli* and humans. Although cytochrome *c* performs the same function in both species, there are differences in the amino acid sequences of the bacterial and human molecules. The number of differences in amino acid sequence provides clues to evolutionary relatedness: fewer differences indicate that organisms are more closely related, diverged relatively recently; many differences indicate that species diverged in the more distant past.

Recently, the study of small-subunit rRNA (SSU rRNA) has gained widespread acceptance as an evolutionary clock molecule. rRNAs are structural components of the ribosomes and are universally conserved. All ribosomes are composed of a large subunit and a small subunit. Prokaryotic cells possess a 70S ribosome (S is for Svedberg units), a measure of size based on centrifugation. The small subunit, 30S, contains about 21 proteins and 16S rRNA (about 1,500 bases long); the large subunit, 50S, contains about 34 proteins and two types of rRNA, the 5S rRNA (120 bases) and the 23S rRNA (about 2,900 bases). Eukaryotic cells, however, have a small subunit of 40S and a large subunit of 60S, but 16S rRNA is also found in the eukaryotic small subunit. Both the 16S and 23S rRNAs contain sufficient information (bases) for reliable phylogenetic analysis, but the 16S rRNA is most widely used in molecular genealogy.

The sequence of the gene coding for 16S rRNA, about 1,500 base pairs (bp) in length, appears to be an excellent evolutionary clock. Mutations in the sequence of 16S rRNA that happened during evolution apparently did not occur uniformly along the length of the gene. Some regions are highly conserved, with no sequence changes; others are far more variable, containing short stretches, or "signature sequences," that are found uniquely in a certain group or groups of organisms. The variable regions are employed in comparative sequence analysis. Hybridization with probes complementary to unique sequences permits the researcher to identify an unknown organism and consign it to the appropriate taxon. For example, the 16S rRNA sequence of humans and chimpanzees is similar, reflecting the recent divergence of hominids and great apes, while the disparate sequence of humans compared with that of bacteria reflects separate evolution over a far longer time span.

As discussed in exercise 22, the phylogenetic system proposed by Carl Woese over 20 years ago indicates that life on earth has evolved along three major pathways: the domains *Bacteria, Archaea,* and *Eucarya.* Many of the data to support this hypothesis come from comparative rRNA sequencing analysis. Recent studies seem to indicate that *Eucarya* is more closely related to *Archaea* than to *Bacteria,* a finding that contradicts previously held notions about these most ancient divergences. Clearly, molecular techniques, especially analysis of conserved genes, provide a modern adjunct to traditional taxonomic tools for elucidating evolutionary relationships.

In this exercise, we are going to use the genomic DNA isolated from organisms representing the five kingdoms in exercise 22. We will use PCR to amplify a specific sequence of the SSU rRNA, then analyze, compare, and contrast the results.

Lab Language

16S rRNA
phylogeny
rRNA
Svedberg unit

Laboratory Safety

Review the sections on physical, biological, and chemical hazards in "Laboratory Safety." Special considerations for this exercise:

boiling agarose
gel electrophoresis
ethidium bromide staining
ultraviolet (UV) transilluminator

Laboratory Sequence

Use DNA samples from exercise 22.

Day 1: Dilution of DNA samples (protocol 1)
PCR amplification of 16S rRNA sequences (protocol 2)

Day 2: Gel electrophoresis and analysis of results

Materials

Class equipment and supplies
10× PCR buffer
Taq DNA polymerase
10 mM deoxynucleoside triphosphates
16S primers: SSU-1 and SSU-2
PCR master mix for 16S rRNA
thermal cycler
microcentrifuge
agarose
10× Tris-borate-EDTA (TBE) buffer
UV transilluminator
ethidium bromide staining solution
Polaroid camera and film

Students will need to prepare
dilutions of genomic DNA samples

1× TBE buffer for electrophoresis
2% agarose gel

Materials per pair of students
7 small PCR tubes of PCR master mix
0.5- to 10-μl micropipette and sterile tips
10- to 100-μl micropipette and sterile tips
100- to 1,000-μl micropipette and sterile tips
marking pen
1 aliquot of loading dye
1 aliquot of PCR oil
gel electrophoresis chamber, gel casting tray, comb
power supply (two pairs of students will share this)

Protocols

PROTOCOL 1 PREPARATION OF GENOMIC DNA FOR PCR

1. Using the genomic DNA isolated in exercise 22, fill in the table below.

Organism	DNA quantity in each lane (μg/μl)	Amount for 500 ng	Amount of 1× TE
Bacteria			
Yeast			
Euglena			
Plant			
Chicken			
Human			

2. Determine how to dilute each sample to a concentration of 1 ng/μl. Prepare 500 μl of 1 ng of each DNA per μl.

3. Document your calculations below.

PROTOCOL 2 AMPLIFICATION OF 16S rRNA SEQUENCES

1. Obtain seven tubes of PCR master mix. Label the tubes and record your key below.

PCR master mix contains all four deoxynucleoside triphosphates, Taq *DNA polymerase, buffer with* Mg^{2+}*, and two different oligonucleotide primers.*

2. Add 20 μl of genomic DNA at 1 ng/μl (isolated in exercise 22) to 30 μl of PCR master mix.

Adding too much DNA to the PCR master mix will inhibit amplification.

3. Add 20 μl of dH_2O to the last tube of PCR master mix as a negative control.

4. Pulse the tubes in the microcentrifuge to mix.

PCR oil is very pure mineral oil: a thin layer prevents evaporation during temperature cycling.

5. Add 50 μl of PCR oil to each tube.

Some thermal cyclers are equipped with a heated lid, which eliminates the need for oil.

6. Place the samples in the thermal cycler.

Programming of the thermal cycler will be demonstrated.

Reaction Conditions for Amplification of 16S rRNA

cycle 1	step 1	93°C for 2 minutes
	step 2	55°C for 1 minute
	step 3	72°C for 1 minute

Cycle 1 contains an extended 93°C step to completely denature template DNA.

cycles 2 to 30	step 1	93°C for 1 minute
	step 2	55°C for 1 minute
	step 3	72°C for 1 minute
cycle 31	step 1	93°C for 1 minute
	step 2	55°C for 1 minute
	step 3	72°C for 2 minutes

Cycle 31 contains an extended elongation step to permit completion of amplification products.

Hold at room temperature or 4°C.

PROTOCOL 3 ANALYSIS OF PCR PRODUCTS

Analysis of PCR products is performed by gel electrophoresis.

2% agarose gels provide optimal resolution of PCR products less than 1,000 bp in size.

1. Prepare a 2% agarose gel. Refer to exercises 7 and 12 for specific details of gel electrophoresis, staining, and photography.
2. Label clean microcentrifuge tubes for each sample. Remove 20 μl of your amplified DNA sample from underneath the oil and pipette it into a clean tube.
3. Add 2 μl of loading dye to each tube and pulse your tubes in the microcentrifuge.
4. Obtain an aliquot of DNA marker in loading dye.

DNA marker sizes must flank 16S rRNA products.

5. Carefully pipette your samples and marker into separate wells of your gel. Record which well contains which sample.
6. Electrophorese your samples until the xylene cyanol (purple dye) in the leading dye is about 2 cm from the end of the gel.

16S rRNA products will be between 500 and 1,500 bp in length.

7. **Wearing gloves,** stain the gel in ethidium bromide solution and photograph the results.

The presence of two bands indicates amplification of contaminating bacterial DNA. Amplification of bacterial 16S rRNA yields a PCR product of ~1,200 bp in length.

References and General Reading

Berschick, P. 1997. One primer pair amplifies small subunit ribosomal DNA from mitochondria, plastids and bacteria. *BioTechniques* **23:**490–494.

Service, R. F. 1997. Biodiversity: microbiologists explore life's rich, hidden kingdoms. *Science* **275:**1740.

Woese, C. 1987. Bacterial evolution. *Microbiol. Rev.* **51:**221–271.

Woese, C., O. Kandler, and M. L. Wheels. 1990. Towards a natural system of organisms: proposal for the domains Archaea, Bacteria, and Eucarya. *Proc. Natl. Acad. Sci. USA* **87:**4576–4579.

Zuckerkandl, E., and L. Pauling. 1965. Molecules as documents of evolutionary history. *J. Theor. Biol.* **8:**357–366.

Discussion

1. How many DNA fragments were amplified in each sample? Explain.
2. Estimate the sizes of the DNA fragments amplified in each sample.
3. Present a hypothesis that explains why a gene with the same function in different organisms has a different sequence length in each.
4. Based on your results, which organisms are most closely related? Explain.
5. Based on your results, which organisms are the least related? Explain.
6. Do your results predicting relatedness compare to what you already know about these organisms based on the five-kingdom system of classification? Explain.

EXERCISE 24

Evolution and Diversity through Protein Homologies

Key Concepts

1. A specific protein from different organisms may have the same function in each but can originate from homologous DNA and amino acid sequences that are not identical.
2. Homologous sequences in DNA and protein are hypothesized to be related by descent from a common ancestral sequence.
3. A protein from different organisms may share antigenic epitopes that can assist in determining the relatedness of the proteins.
4. Organisms that are phylogenetically most diverse from one another are least likely to share protein homologies.

Goals

By the end of this exercise, you will be able to

- discuss sequence homology as a tool in phylogenetic studies
- explain the differences between antigen and epitope
- describe how antibodies may be used to determine relatedness of different organisms

Introduction

DNA and consequently RNA and protein molecules evolve. Scientists believe that each protein family may have descended from a single ancestral gene. Over time, mutations, especially duplications, followed by divergence and additional mutations, may have given rise to related proteins possessing new functions.

The study of DNA nucleotide sequences and protein amino acid has become a most important tool for molecular systematists. The analysis of similarities and differences allows the researcher to determine relationships within various taxa: genera, species, and even races, tribes, and subspecies. Molecular comparison establishes phylogenetic connections and enables us to infer common ancestry among taxa of living organisms. Molecular systematists believe that the greater the similarities between the amino acid sequences of a protein found in two or more different organisms, the closer their relationship. Inversely, the greater the differences, the more distant the relationship.

One well-studied example is the globin protein family. Comparison of protein sequences indicates that myoglobin (oxygen carrier molecule in muscle) and the hemoglobins (oxygen carrier molecules in blood) are related to each other and evolved from a common ancestral globin. Later, the hemoglobin branch diverged to yield α- and β-globins as well as embryonic and fetal chains within the α- and β-globins. A comparison of the β-chain of hemoglobin from gorilla, mouse, and frog with the β-chain of human hemoglobin reveals 1, 27, and 67 amino acid differences, respectively.

Amylase

Amylases are hydrolytic enzymes that catalyze the breakdown of the polysaccharide starch (amylose and amylopectin) into the disaccharide maltose. Two types of amylases have been identified in different organisms: α-amylases and β-amylases.

α-Amylases degrade starch, creating maltose, an α-1,4-glycosidic linkage configuration. α-Amylases are present in all living organisms; however, the enzymes vary remarkably between species and even from tissue to tissue within a single species. In mammals, two principal α-amylases are found, salivary amylase and pancreatic amylase. These isoenzymes are very similar, each with a molecular mass between 50,000 and 55,000 Da (daltons), and both break down starch. Nevertheless, the two enzymes differ in electrophoretic mobility. Other varieties of α-amylase are found in bacteria, yeasts, molds, and plants. Microbially produced α-amylases have great value in industrial applications. The enzymes are used to begin production of high fructose corn syrup, as an additive in laundry detergent, and to degrade sizing in the textile and paper industries.

β-Amylases are commonly found in grains such as barley, corn, and wheat, as well as in sweet potatoes and yams. They break down starch into β-maltose. Another difference in the activity of α- and β-amylases is that β-amylase cannot completely degrade amylopectin, because it lacks the ability to break down amylopectin molecular branching. For complete degradation of starch into maltose, both enzymes are necessary. For further breakdown of starch into glucose, a third enzyme, glucoamylase (glucosidase), is required.

It is hypothesized that all amylases evolved from a common ancestral amylase gene, which diverged first into α- and β-amylase genes. Later, these two genes branched further, giving rise to different tissue-specific isoenzyme genes. The evolution of the amylase gene family seems to be due to gene duplication. Evidence of this evolutionary pattern is shown in *Bacillus polymyxa*, which carries a 3,588-kb amylase gene that codes for both α- and β-amylase. *B. polymyxa* appears to be one of the few species that have a single gene for both enzymes. This gene could be closely related to the ancestral amylase gene.

Homologies between bacterial and eukaryotic α-amylase have also been reported, as well as homologies between bacterial and eukaryotic β-amylase. For example, the β-amylase found in *Bacillus cereus* (514 amino acid residues) contains 10 well-conserved regions found between the N terminus and the 430-residue region. The C-terminal region of 90 residues, though, has no similarity with that of the plant β-amylases. Thus, *B. cereus* β-amylase has a homology of 52.7% with the β-amylase of *B. polymyxa*, 43.4% with *Clostridium thermosulfurogenes*, 31.8% with *Arabidopsis thaliana*, 31.5% with barley, 29.9% with sweet potato, and 28.9% with soybean.

For this exercise, we will compare the enzyme amylase from organisms representing the five kingdoms: bacteria (*Escherichia coli*), yeasts (*Saccharomyces cerevisiae*), plant tubers and seeds, human cheek cells, and mouse pancreas, using sodium dodecyl sulfate-polyacrylamide gel electrophoresis (SDS-PAGE) and Western blotting as performed in exercises 18 and 19.

Lab Language

α-amylase
amylase
β-amylase
electrophoretic mobility
homology
isoenzyme
phylogeny
taxonomy

Laboratory Safety

Review the sections on physical, biological, and chemical hazards in "Laboratory Safety." Special considerations for this exercise:

use of human cheek cells
use of microbes
heating/hot samples
gel electrophoresis
SDS-PAGE: acrylamide, ammonium persulfate, *N,N,N′,N′*-tetramethylethylenediamine (TEMED), dithiothreitol (DTT)
immunodetection: 5-bromo-4-chloro-3-indolylphosphate (BCIP), nitroblue tetrazolium

Laboratory Sequence

The class will be divided into three groups, each testing a different antibody. Each group of students will be responsible for preparing two blots. The immunodetection results will be shared between all groups to reach a conclusion regarding amylase evolution.

Day 1: Preparation of samples
Preparation of SDS-polyacrylamide gel
Gel electrophoresis
Gel staining
Western blotting

Day 2: Drying of stained gel
Immunodetection and developing

Materials

SDS-PAGE

Class equipment and supplies

30% acrylamide-bisacrylamide, 37.5:1
1.5 M Tris, pH 8.8
10% SDS
10% ammonium persulfate
TEMED
1.0 M Tris, pH 6.8
Pasteur pipettes and bulbs
clinical centrifuge
microcentrifuge
heat block or boiling water bath
2× Laemmli sample buffer (LSB)
protein marker in 2× LSB
10× SDS running buffer
Coomassie blue stain
destain
gel-drying apparatus
dH_2O

Students will need to prepare

10% polyacrylamide gel
5% stacking gel
300-ml 1× SDS running buffer

Materials per pair of students

1-cm-thick slice of potato
1-cm-thick slice of sweet potato (*Ipomoea batatas*)
1-cm-thick slice of sweet yam
6 soaked barley grains
overnight yeast culture
overnight culture of *Bacillus subtilis* or *Bacillus amyloliquefaciens*
0.9% sterile sodium chloride
15-ml conical centrifuge tube
mouse pancreas
vertical gel electrophoresis system with glass plates, spacers, comb, and chamber
gloves
microcentrifuge tubes
1 aliquot of 2× LSB
0.5- to 10-μl micropipette and sterile tips
100- to 1,000-μl micropipette and sterile tips
power supply capable of 250 volts (shared by 2 to 4 groups)
500-ml graduated cylinder
50-ml graduated cylinder
small, cold mortar and pestle

Western Blot

Class equipment and supplies

10× Tris-glycine
methanol

blotting paper or paper towels
Whatman 3MM paper

Materials per pair of students
tile support and container for blotting
gloves

Students will need to prepare
300 ml of blotting buffer
3 wicks approximately 10 by 30 cm
nitrocellulose cut to gel size
3 layers of Whatman paper cut to gel size

Immunodetection

Class equipment and supplies
NaCl
KCl
Na_2HPO_4
KH_2PO_4
Tween 20
dried milk
primary antibody (1° Ab)
- anti-human α-amylase (salivary or pancreatic amylase)
- anti-bacterial amylase
- anti-β-amylase (sweet potato)

secondary antibody (2° Ab) conjugated to alkaline phosphatase
25× developer concentrate
nitroblue tetrazolium solution
BCIP solution
shaking platform (optional)

Materials per pair of students
0.5- to 10-μl micropipette and sterile tips
10- to 100-μl micropipette and sterile tips
container for incubations

Students will need to prepare
500 ml of physiologically buffered saline with 0.5% Tween 20 (PBS-T)
25 ml of PBS-T with 5% dried milk
15 ml of 1:1,000 1° Ab in PBS-T
15 ml of 1:1,000 2° Ab in PBS-T
25-ml developer solution

Protocols

PROTOCOL 1 PREPARATION OF SWEET POTATO, YAM, POTATO, AND BARLEY EXTRACTS

1. Cut the tissue (potato, yam, sweet potato) into a 1- by 1- by 1-cm cube.
2. Grind one cube of tissue or the barley grains into a paste using a small cold mortar and pestle.
3. Place the paste into a 1.5-ml test tube. Label the tube.
4. Add 500 μl of 2× LSB.

2× LSB contains SDS, which gives all proteins a net negative charge, and DTT, which breaks disulfide bonds.

5. Vortex well to mix.
6. Incubate the samples at room temperature for 5 minutes.
7. Centrifuge at 14,000 rpm for 3 minutes to pellet the tissue remnants.

Chunks of tissue will interfere with electrophoresis.

8. Remove the supernatant to a clean, labeled microcentrifuge tube and use for SDS-PAGE analysis.
9. Heat the samples at 100°C for 5 minutes, *just before loading.*

Heat assists with protein denaturation.

PROTOCOL 2 PREPARATION OF BACTERIAL AND YEAST CELL EXTRACTS

1. Obtain an overnight broth culture of *E. coli* and of *S. cerevisiae.*
2. Label two microcentrifuge tubes. Transfer approximately 1.0 ml of each culture into the appropriate tube, using a Pasteur pipette.

3. Pulse the tubes in the microcentrifuge for 5 to 10 seconds to pellet the microbes.

4. Carefully remove most of the supernatant by pouring it into a container of disinfectant.

5. Add 500 µl of 2× LSB to the pellet in each tube.

2× LSB contains SDS, which gives all proteins a net negative charge, and DTT, which breaks disulfide bonds.

6. Vortex each tube to lyse the cells.

7. Use the resulting liquid for SDS-PAGE analysis.

8. Heat the samples at 100°C for 5 minutes, *just before loading.*

Heat assists with protein denaturation.

PROTOCOL 3 PREPARATION OF MOUSE PANCREAS EXTRACT

1. Obtain a 1- by 1- by 1-cm cube of frozen pancreas.

2. Using a chilled mortar and pestle, and working on ice, pulverize the sample.

3. Transfer the mashed tissue into a labeled microcentrifuge tube.

4. Add 500 µl of 2× LSB.

2× LSB contains SDS, which gives all proteins a net negative charge, and DTT, which breaks disulfide bonds.

5. Vortex to mix.

6. Let the samples incubate at room temperature for 5 minutes.

7. Centrifuge at 14,000 rpm for 3 minutes to pellet the tissue remnants.

Chunks of tissue will interfere with electrophoresis.

8. Remove the supernatant to a clean, labeled microcentrifuge tube and use for SDS-PAGE analysis.

9. Heat the samples at 100°C for 5 minutes, *just before loading.*

Heat assists with protein denaturation.

PROTOCOL 4 PREPARATION OF CHEEK CELL EXTRACT

1. Rinse your mouth well with tap water.

2. Obtain cheek cell pellet as described in exercise 10, protocol 1.

3. Add 1 ml of 2× LSB to the cell pellet.

2× LSB contains SDS, which gives all proteins a net negative charge, and DTT, which breaks disulfide bonds.

4. Pipette up and down to resuspend and lyse the cells.

5. Transfer the entire sample to a 1.5-ml microcentrifuge tube. Use this sample for SDS-PAGE analysis.

6. Heat the samples at 100°C for 5 minutes, *just before loading.*

Heat assists with protein denaturation.

PROTOCOL 5 SDS-PAGE: 10% POLYACRYLAMIDE GELS

PROTOCOL 6 COOMASSIE BLUE STAINING

PROTOCOL 7 WESTERN BLOT

PROTOCOL 8 IMMUNODETECTION USING ABOVE-MENTIONED ANTIBODIES

See exercises 18 and 19.

References and General Reading

Gumucio, D. L., K. Wiebauer, A. Dranginis, L. C. Samuelson, L. O. Treisman, R. M. Caldwell, T. K. Antonucci, and M. H. Heisler. 1985. Evolution of the amylase multigene family. *J. Biol. Chem.* **260:**13483–13489.

Hagenbuchle, O., R. Bovey, and R. A. Young. 1980. Tissue-specific expression of mouse alpha-amylase genes: nucleotide sequence of isoenzyme mRNAs from pancreas and salivary glands. *Cell* **21:**179–187.

Madigan, M. T., J. M. Martinko, and J. Parker. 1997. *Brock Biology of Microorganisms,* 8th ed. Prentice-Hall, Inc., Upper Saddle River, N.J.

Nanmori, T., M. Nagai, Y. Shimizu, R. Shinke, and B. Mikami. 1993. Cloning of the beta-amylase gene from *Bacillus cereus* and characteristics of the primary structure of the enzyme. *Appl. Environ. Microbiol.* **59:**623–627.

Uozumi, N., K. Sakurai, T. Sasaki, S. Takekawa, H. Yamagata, N. Tsukagoshi, and S. Udaka. 1989. A single gene directs synthesis of a precursor protein with beta- and alpha-amylase activities in *Bacillus polymyxa. J. Bacteriol.* **171:**375–382.

Worthington Biochemical Corporation. 1993. *Worthington Enzyme Manual.* Worthington Biochemical Corporation, Freehold, N.J.

Discussion

1. List similarities and differences obtained with each antibody-probed membrane.
2. Which antibodies bind to amylases present in all blots? Explain.
3. Which antibodies bind only to specific amylases? Explain.
4. Based on your results, predict the relatedness of the organisms used in this experiment. Discuss your answer.
5. Do your results predicting relatedness confirm or dispute what you already know about these organisms based on the five-kingdom system of classification? Explain.
6. Draw a scheme of the evolution of the amylases.
7. What is the purpose of rinsing your mouth with water before collecting cheek cells?

SECTION VIII

Internet- and Computer-Assisted Gene Analysis

EXERCISE 25

Internet Gene Bank Search

Key Concepts

1. Bioinformatics is the branch of science devoted to computational information management systems used to collect, store, analyze, and disseminate biological information.
2. Many DNA and protein sequence databases are readily available through the World Wide Web.
3. Important gateway websites, containing links to sequence databases and computational centers for sequence analysis, include the National Center for Biotechnology Information (NCBI) (http://www.ncbi.nlm.nih.gov/), the European Bioinformatics Institute (EBI) (http://www.ebu.ac.uk/), and Pedro's BioMolecular Research Tools (Pedro's) (http://www.fmi.ch/biology/research_tools.html).
4. GenBank is a sequence database maintained by the National Center for Biotechnology Information, the National Library of Medicine, and the National Institutes of Health.

Goals

By the end of this exercise, you will be able to

- access NCBI, EBI, and Pedro's websites
- describe databases and websites available for sequence analysis
- access and locate sequences available in GenBank

Introduction

Bioinformatics

Bioinformatics is a rapidly expanding field of biological research. Born of the union between information technology and gene-sequencing projects, bioinformatics may be defined as the discipline concerned with the computational systems used to acquire, store, analyze, disseminate, and publish biological information. The need for qualified, well-trained personnel to work in bioinformatics is currently unmet. Job vacancies seem to be abundant and are found in a variety of academic and industrial settings. "Bioinformaticists" may be people who know molecular biology and sequence analysis techniques and who can validly interpret results of these applications. Others write software or design and build databases to handle complex analytical functions. Still others build complex systems and integrate them electronically with servers in other parts of the country or the world. Pharmaceutical companies hire bioinformatics specialists to assist in drug discovery and development through analysis of genomic data.

Several colleges and universities, including the Keck Center (a consortium among Carnegie Mellon University, the University of Pittsburgh, and the Pittsburgh Supercomputing Center), Rensselaer Polytechnic Institute, the University of California at Berkeley, Rutgers University, and others offer advanced degree programs in bioinformatics as well as stand-alone courses. Stanford University offers a series of three courses in bioinformatics online. Students with a dual interest in molecular biology and computer science will have many opportunities for education and employment in this rapidly growing field.

Databases for nucleic acids and proteins, and analytical tools to manage them, are available to researchers, students, and the public on the World Wide Web at no cost. Other software systems for sequencing and analyzing nucleic acid and protein data are commercially available. Some examples include those produced by DNAStar, the Oxford Molecular Group, and Perkin-Elmer/Applied Biosystems. Users purchase a single copy of the proprietary software or a site license to use the software at multiple workstations. Virtually all such programs can interface with the publicly available data archives accessible on the Internet.

Bioinformatics, a new young field, is in a state of flux: the only constant element is change.

Databases and Data Analysis

Methods for determining DNA sequence were first described in the 1970s (see exercise 28). Since then, a wealth of sequence information has been obtained and deposited in several essential centralized locations. These generalized databases include GenBank (Fig. 25.1), the European Molecular Biology Laboratory (EMBL), and the DNA Database of Japan (DDBJ).

Other, more specialized databases have been constructed by groups with an interest in a particular group of molecules and sequences or in the genomics of a particular species. Examples of these include the Transcription Factor Database at the National Institutes of Health, the Genetic and Molecular Database for *Drosophila* (FlyBase) at Harvard University, the Database of Major Histocompatibility Complex Binding Peptides maintained at the Walter and Eliza Hall Institute for Medical Research in Melbourne, Australia, and Mouse Genome Informatics at the Jackson Laboratory in Maine.

Databases and database analysis tools allow a researcher to probe for a desired sequence. If an investigator wishes to study a particular gene previously isolated and sequenced, then he or she may obtain the sequence from the database. This eliminates the need to reisolate and resequence the gene of interest.

A researcher who believes that he or she has isolated a new gene may be able to discover information about the gene sequence from a database. Therefore, isolating the full gene or completely sequencing it may not be necessary. Partial sequence information may be compared with existing sequences in a database to probe for homologs. A researcher may discover that the several hundred nucleotides sequenced exactly match another sequence previously deposited in the database. If this is the case, he or she may collaborate with the person who originally deposited the sequence, rather than going through the laborious task of sequencing the whole gene a second time. If a probe of the database shows no matches, then the researcher has identified a new gene, or one that has not yet been deposited in the database. Sometimes, a probe of a database with sequence information yields similar but not identical sequences. This may indicate that the particular gene under study is homologous to that of another species, or is homologous to other genes within the same species, possibly indicating the presence of a new member in a family of genes.

In this exercise, you will be introduced to some of the available databases and World Wide Web sites for sequence analysis. Several sites provide "gateway" facilities in that they are major repositories of archived data and provide links to databases and database analysis tools around the world. These include NCBI, EBI, and Pedro's. GenBank, EMBL, and DDBJ share and update sequence data daily. Pedro's was last updated on June 16, 1995 (as of this writing). Since many

National Center for Biotechnology Information

National Library of Medicine   National Institutes of Health

PubMed   Entrez   BLAST   OMIM   Taxonomy   Structure

Search GenBank for [ ] Go

SITE MAP

About NCBI
general and contact information

GenBank
sequence submission support and software

Molecular databases
sequences, structures and taxonomy

Literature databases
PubMed and OMIM

Genomic biology
whole genomes and related resources

Tools
for data mining

Research at NCBI
people, projects and seminars

Education
teaching resources and on-line tutorials

FTP site
download data and software

What does NCBI do?

Established in 1988 as a national resource for molecular biology information, NCBI creates public databases, conducts research in computational biology, develops software tools for analyzing genome data, and disseminates biomedical information - all for the better understanding of molecular processes affecting human health and disease.

COGs expanded   Archives

Clusters of Orthologous Groups (COGs) now incorporates 21 complete genomes, tripling the number of organisms represented. COGs can now be searched by full text, protein/gene name as well as phylogenetic patterns. On-line help is now available.

NCBI in the News

PubMed was described as "an outstanding, free service to the research and education communities", and OMIM was named "*the* authoritative reference for information on the inheritance of human characteristics" in the August, 1999, issue of the American Library Association magazine, CHOICE.

Disclaimer   Privacy statement

Revised October 13, 1999

Hot Spots

- Cancer genome anatomy project
- Clusters of orthologous groups
- Coffee Break
- Electronic PCR
- Genes and disease
- Human genome resources
- Human/mouse homology maps
- LocusLink
- Malaria genetics & genomics
- ORF finder
- Reference sequence project
- Retrovirus resources
- Serial analysis of gene expression
- UniGene
- VecScreen

Figure 25.1 NCBI home page.

Internet sites may have a half-life of four years, a few links may be outdated or extinct.

You will search GenBank for the complementary DNA (cDNA) sequence for human growth hormone, and then, with your acquired bioinformatics navigational skills, search for any gene of interest to you.

Lab Language

bioinformatics
database
EBI
GenBank
NCBI
Pedro's BioMolecular Research Tools
URL

Materials

Materials per pair of students

computer with Internet access
Web browser (such as Netscape or Internet Explorer)
printer

Protocols

PROTOCOL 1 MOLECULAR BIOLOGY DATABASES AVAILABLE ON THE WORLD WIDE WEB

1. Locate Pedro's BioMolecular Research Tools (http://www.fmi.ch/biology/research_tools.html).

Selecting a link allows you to navigate to other Internet sites.

2. Select "Part 1: Molecular Biology Search and Analysis." Nearly 200 links are listed that provide access to various databases and sites for analyzing sequences.

URL stands for Uniform Resource Locator.

The meta-search engine at the Toolkit home page (http://toolkit.imsa.edu) can help you locate database web sites.

3. Identify three different links, list the URL of the site, and describe the type of database or sequence analysis available at that site.
4. Locate the NCBI home page (http://www.ncbi.nlm.nih.gov/).
5. Select "Genes and Disease."
6. How many major disease categories are listed?
7. How could you locate a disease associated with a specific chromosome?
8. Locate the EBI home page (http://www.ebi.ac.uk/).
9. Select "Search EBI" in the left column.
10. Follow the instructions, and enter your search query.
11. How many documents were retrieved?

PROTOCOL 2 SEARCH GENBANK FOR HUMAN GROWTH HORMONE

Web pages are continually updated; the NCBI home page may change.

If the NCBI home page is different from that illustrated, carefully read and follow the new instructions.

1. Locate the NCBI home page (http://www.ncbi.nlm.nih.gov/).
2. Select "Search GenBank" for "human growth hormone" at the top of the page (Fig. 25.1).
3. Select "Go."

GenBank may also be located through Pedro's BioMolecular Research Tools.

4. How many documents were retrieved?

You can also search a protein sequence database.

5. In the "Add Term(s) to Query" search box, type in "cDNA" and then select "Search" (Fig. 25.2).

Adding cDNA will narrow the search parameters.

6. How many documents were retrieved?
7. Select "Retrieve [number] Documents."
8. Of the documents retrieved, what is the accession number that contains the cDNA sequence for the human growth hormone?

Other report styles are also available from the site.

9. Select the document with the correct accession number and click on "Display GenBank report."
10. Click "Save" and follow the prompts from your computer.

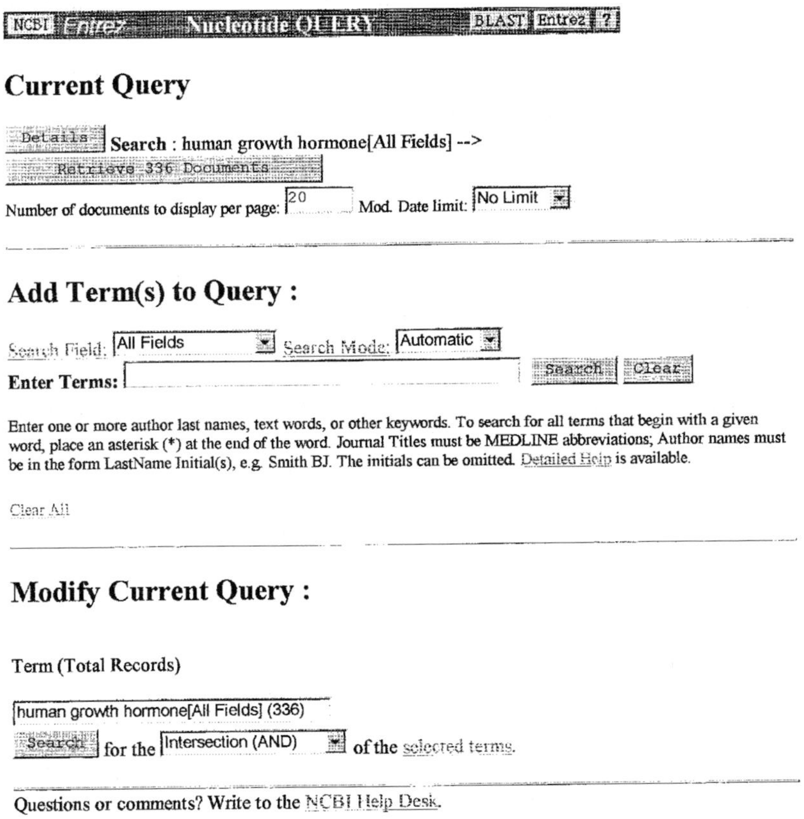

NCBI Entrez Nucleotide QUERY BLAST Entrez ?

Current Query

Details Search : human growth hormone[All Fields] -->
Retrieve 336 Documents

Number of documents to display per page: 20 Mod. Date limit: No Limit

Add Term(s) to Query :

Search Field: All Fields Search Mode: Automatic
Enter Terms: Search Clear

Enter one or more author last names, text words, or other keywords. To search for all terms that begin with a given word, place an asterisk (*) at the end of the word. Journal Titles must be MEDLINE abbreviations; Author names must be in the form LastName Initial(s), e.g. Smith BJ. The initials can be omitted. Detailed Help is available.

Clear All

Modify Current Query :

Term (Total Records)

human growth hormone[All Fields] (336)

Search for the Intersection (AND) of the selected terms.

Questions or comments? Write to the NCBI Help Desk.

Figure 25.2 GenBank modified search page.

11. What are some of the other sequences obtained with a search for the cDNA nucleotide sequence of the human growth hormone?

PROTOCOL 3 USING WWW ENTREZ TO SEARCH FOR HUMAN GROWTH HORMONE

1. Locate the GenBank database (http://ncbi.nlm.nih.gov/).

Websites are continually updated; follow current instructions.

2. Select "Entrez" at the top of the NCBI home page.

WWW Entrez searches GenBank and other linked sequence depositories.

3. Select "Click here to enter the new Entrez System."
4. Select "Nucleotide" and enter "growth hormone." Select "Go" (Fig. 25.3).

This will allow you to search the DNA sequence database.

5. How many documents were retrieved?
6. Select "Clear."
7. Select "Nucleotide" and enter "human growth hormone." Select "Go."
8. How many documents were retrieved?
9. Select "Clear."
10. Select "Nucleotide" and enter "human growth hormone cDNA." Select "Go."
11. How many documents were retrieved?

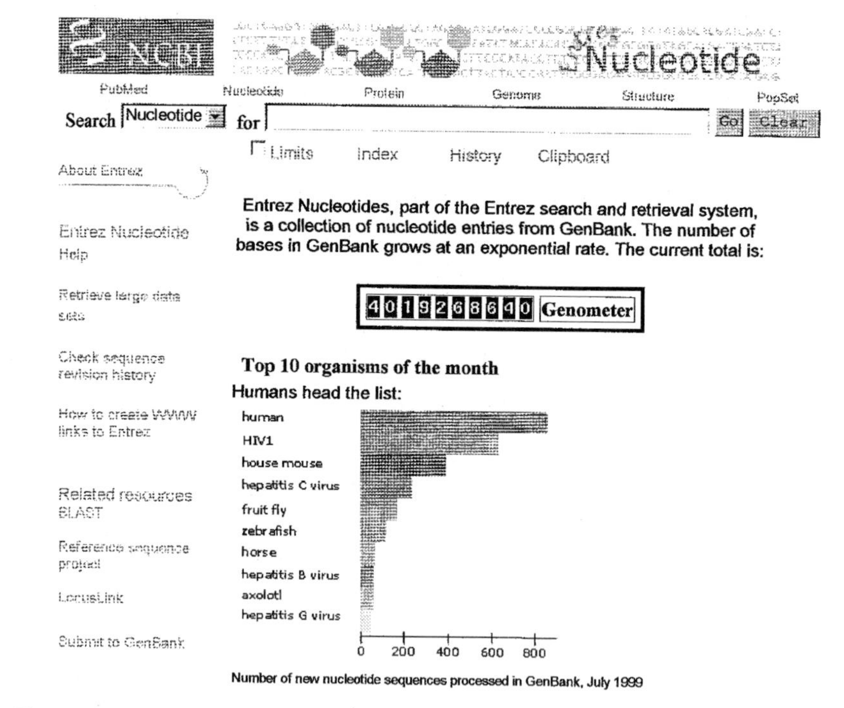

Figure 25.3 Entrez nucleotide search page.

12. Select "Protein" and enter "growth hormone." Select "Go" (Fig. 25.4).

13. How many documents were retrieved?

14. Select "Clear."

15. Select "Protein" and enter "human growth hormone." Select "Go."

16. How many documents were retrieved?

PROTOCOL 4 SEARCH GENBANK FOR SEQUENCE OF INTEREST

1. Select a gene that you would like to know more about. Before you log on, prepare your search strategy.

2. In which species or organism type would you like to look for this sequence?

3. Is the sequence you desire DNA, RNA, or protein?

4. If you are searching for a DNA sequence, is the sequence cDNA or genomic DNA?

5. If protein, what do you know about it?

6. Access GenBank as in protocol 2, steps 1 and 2, or WWW Entrez as in protocol 3, steps 1 and 2.

7. Choose the appropriate database.

8. Enter the key terms for searching and select "Go."

9. How many documents were retrieved?

10. Do you need to narrow your search parameters? How can this be done?

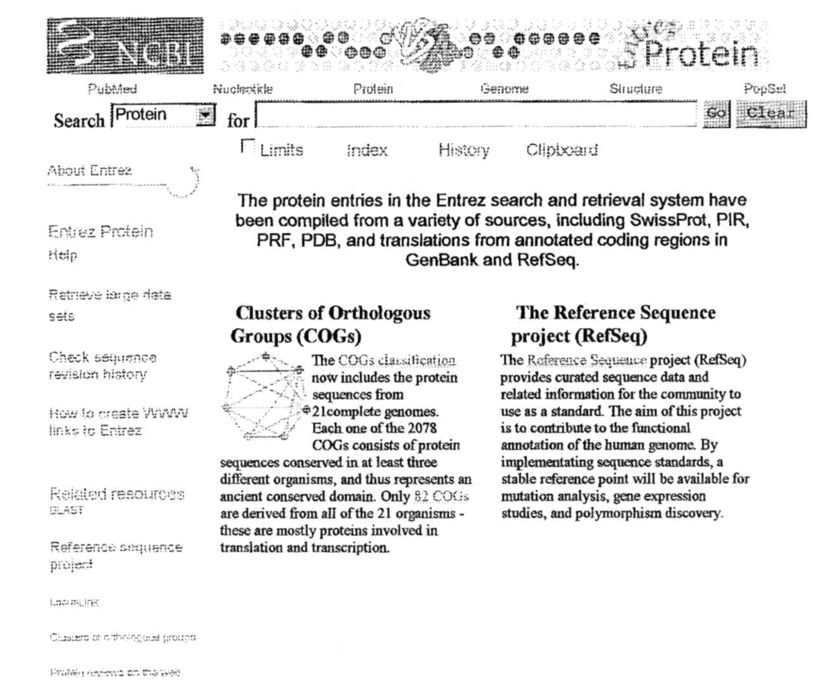

Figure 25.4 Entrez protein search page.

11. What is the accession number of the sequence that you were searching for?

12. List which publication this information is from.

13. List three pieces of data available about this gene from the search.

References and General Reading

Gelbart, W. M. 1998. Databases in genomic research. *Science* **282:**659–661.

Gershon, D. 1997. Bioinformatics in a post-genomics age. *Nature* **389:**417–418.

Rogers, R. S. 1998. Bioinformatics begins to buzz. *Chem. Eng. News* **76:**29–33.

Sobral, B. W. S. 1997. Common language of bioinformatics. *Nature* **389:**418–420.

Internet Sites

http://www.ebi.ac.uk/ (European Bioinformatics Institute [EBI])
http://www.fmi.ch/biology/research_tools.html (Pedro's BioMolecular Research Tools)
http://www.ncbi.nlm.nih.gov/ (National Center for Biotechnology Information [NCBI])

Discussion

1. What is GenBank? What is its purpose?
2. What is Pedro's BioMolecular Tools? NCBI? EBI? What purpose do these sites serve?
3. Why did adding the key term "cDNA" to your search for the sequence of the human growth hormone result in retrieval of fewer documents?
4. How could a search be performed that accesses *only* the desired sequence?
5. Which sequence did you choose for your GenBank search and why?
6. Was your search successful? Why or why not?
7. What essential terms or search parameter did you find most useful in narrowing your search?
8. Why might you obtain different search results from one time to another?

EXERCISE 26

Computer-Assisted Sequence Analysis

Key Concepts

1. Webcutter is a set of restriction enzyme analysis tools on the World Wide Web that can predict restriction enzyme cut sites.
2. ExPASy is a World Wide Web molecular biology server that provides tools to analyze protein and DNA sequences.
3. BLAST (Basic Local Alignment Search Tool) at the National Center for Biotechnology Information (NCBI) is a similarity search program, which enables a researcher to probe databases for identical or homologous sequences.

Goals

By the end of this exercise, you will be able to

- use Webcutter to generate a restriction enzyme map of a DNA sequence
- perform a BLAST search for homologous sequences
- use the ExPASy server to translate DNA sequence into amino acid sequence
- determine correct reading frame

Introduction

A major goal of bioinformatics is to provide the tools required to analyze the rapidly increasing nucleic acid and protein data sets archived in globally available repositories. Some workers estimate that the doubling time for sequence information deposited in the major archives, such as GenBank, the European Molecular Biology Laboratory (EMBL), the DNA Database of Japan (DDBJ), and SWISS-PROT, may be as short as 12 months! To deal with this rapidly expanding information load, bioinformatics specialists have devised, and continue to devise, computational tools capable of archiving, searching, analyzing, and integrating available data. A researcher may use a variety of strategies to compare his or her results with those previously deposited.

Many analysis tools are available in the public domain, easily accessible through the World Wide Web. Access to tools is also available at selected file transfer protocol (FTP) servers, e-mail servers, and Gopher servers. Educators and students with easy, open access to the Internet often find it more convenient to use World Wide Web sites.

Although proprietary, commercial sequence analysis tools are available for purchase, this exercise will use programs and servers that are publicly available.

Webcutter

Once a gene has been isolated and its nucleotide sequence has been determined, researchers use a variety of tools to begin further characterization. Mapping restriction enzyme cleavage sites (see exercise 7 for a review of restriction enzymes) can provide an investigator with the means to devise cloning methods or truncate a gene for protein studies.

Important information includes whether a particular restriction enzyme site appears only once, is present in the middle of a coding sequence, or appears at either the 5′ or 3′ end of a gene. A few restriction enzymes recognize different sequences but yield the same cohesive ends, useful information for designing cloning strategies. It may be critical to know whether a particular restriction enzyme cuts a DNA sequence at all. Over 2,750 restriction enzymes have been identified with 211 different specificities. It is an enormous job to manually predict where each restriction enzyme would cut a particular DNA fragment.

Computer programs have been developed to determine which restriction enzyme recognition sites exist in a given DNA sequence and which enzymes will not cut it at all. One of these is Webcutter (http://www.firstmarket.com/cutter) (Fig. 26.1). An individual can either enter a DNA sequence manually, cut a sequence from a word-processing document and paste it into the site, or retrieve a sequence directly from GenBank. The program will determine restriction enzyme cleavage sites. Since many different restriction enzymes are available, a worker may instruct the program to predict recognition sites for selected enzymes. Additionally, a researcher may wish to obtain information about the restriction enzymes that cut the sequence a limited number of times or not at all. Other desirable informa-

Figure 26.1 Webcutter home page.

Welcome to Webcutter 2.0! This new version of Webcutter is a complete rewrite. Along with cleaner and more maintainable code, I am pleased to introduce the following new features:

- **Rainbow cutters** Highlight your favorite enzymes in color or boldface for easy at-a-glance identification
- **Silent cutters** Find sites which may be introduced by silent mutagenesis of your coding sequence
- **Sequence uploads** Input sequences directly into Webcutter from a file on your hard drive without needing to cut-and-paste
- **Degenerate sequences** Analyze restriction maps of sequences containing ambiguous nucleotides like N, Y, and R.
- **Circular sequences** Choose whether to treat your sequence as linear or circular
- **Enzyme info** Click into the wealth of references and ordering information at New England BioLabs' REBASE, directly from your restriction map results

Plus all of the features Webcutter has always had, from automatic sequence search-and-entry from NCBI's GenBank to its easy customizable interface and clean simple results format. For a mini-manual on how Webcutter 2.0 works, how to get the most from it, and some of its known limitations, please click here.

There are three ways to input your sequence:

1. You can copy-and-paste it or type it into the box below
2. You can upload a sequence file from your computer by clicking the "Browse..." button below. If you are not using Netscape, you may not see the browse button and cannot use this feature (sorry).

Upload Sequence File

3. Or, you can search NCBI's GenBank for your sequence by entering search terms below. Webcutter will submit your search to GenBank and automatically retrieve and enter your sequence for you.

Query GenBank

For more advanced searches, please visit one of these sites
- The National Center for Biotechnology Information's GenBank
- European Bioinformatics Institute

tion might include enzymes with recognition sites of a given base-pair length. These capabilities are some of Webcutter's features.

Searches Using BLAST

Because the DNA sequences isolated by various researchers are rapidly expanding, someone else may have previously identified the same sequence. This does not mean that one's work is redundant; the same sequence may be isolated from two or more different tissue sources or organisms, which may indicate widespread use of a particular gene. However, if someone has previously sequenced the entire gene of interest, it is usually unnecessary to sequence it a second or third time.

BLAST provides a way to probe sequence databases, such as GenBank, EMBL, DDBJ, and others, for identical or homologous sequences. If a similarity search results in 100% identity, then the gene in question is probably the same as the one previously identified and deposited. BLAST search programs can also identify sequences that are similar, but not identical, and assign an expected value to determine the number of hits expected purely by chance. The researcher then can assess whether a gene homologous across several species may have been identified. Additionally, a similarity search might indicate that the newly isolated gene may be part of a larger gene family, with similar sequences and functions. Several types and levels of BLAST searches are accessible at the website.

The ExPASy Molecular Biology Server

Upon isolation of a DNA fragment and subsequent determination of nucleotide sequence, a researcher needs to know whether he or she has isolated the entire gene sequence and needs to analyze the amino acid sequence of the translated protein. Since DNA is double stranded, one strand is the coding strand and the other is the complementary strand. This means that a researcher must examine six reading frames to determine which strand and triplet frame is used for transcription and translation. Computer analysis will translate both DNA strands and the three possible reading frames of each. An ExPASy computerized program tool quickly translates all six possible reading frames, indicating potential methionine (Met) start residues and stop codons.

In eukaryotes, the first amino acid in a protein is a Met residue, encoded by an ATG nucleotide triplet. Finding an ATG codon/Met residue potentially can indicate the beginning of a gene/protein, although Met residues may be distributed throughout a protein sequence. The next step is to determine whether there are any stop codons in frame with the ATG start codon. The stop codons are TAA, TAG, and TGA. There will be no stop codon in frame in the correct coding sequence.

Performing this type of sequence analysis manually is excruciatingly time-consuming. ExPASy permits rapid analysis of translation predictions. Other proteomics tools within the ExPASy suite of programs include protein structure analysis and prediction. The ExPASy home page also contains a variety of links to other molecular biology resources and servers.

In this exercise, you will submit a sequence to Webcutter, GenBank, and ExPASy for generation of a restriction enzyme map, analysis and similarity searching of the gene sequence, determination of amino acid sequence, and evaluation of the resulting reading frames for correctness.

Lab Language

BLAST
ExPASy
Webcutter

Materials

Materials per pair of students
computer
Internet access
Web browser (such as Netscape or Internet Explorer)
printer
floppy disks (optional)

Protocols

PROTOCOL 1 GENERATION OF A RESTRICTION ENZYME MAP

This sequence will be used again in protocols 2 and 3.

Be sure to proofread your entry before continuing.

1. Enter the following sequence into a word-processing document and save it in an appropriate location on the hard drive, floppy disk, zip drive, desktop, etc.:

```
1     tgtgttcact agcaacctca aacagacacc atggtgcacc
41    tgactcctga ggagaagtct gccgttactg ccctgtgggg
81    caaggtgaac gtggatgaag ttggtggtga ggccctgggc
121   aggctgctgg tggtctaccc ttggacccag aggttccttg
161   agtcctttgg ggatctgtcc actcctgatg ctgttatggg
201   caaccctaag gtgaaggctc atggcaagaa agtgctcggt
241   gcctttagtg atggcctggc tcacctggac aacctcaagg
281   gcacctttgc cacactgagt gagctgcact gtgacaagct
321   gcacgtggat cctgagaact tcaggctcct gggcaacgtg
361   ctggtctgtg tgctggccca tcactttggc aaagaattca
401   ccccaccagt gcaggctgcc tatcagaaag tggtggctgg
441   tgtggctaat gccctggccc acaagtatca ctaagctcgc
481   tttcttgctg tccaatttct attaaaggtt cctttgttcc
521   ctaagtccaa ctactaaact gggggatatt atgaagggcc
561   tt
```

Websites are continually updated; if the Webcutter home page is different from that illustrated, follow the new directions.

2. Locate Webcutter (http://www.firstmarket.com/cutter).

3. Scroll to the bottom of the Webcutter news page. Select use of the latest version of Webcutter.

4. Scroll down the Webcutter 2.0 page and cut and paste your sequence into the sequence box (Fig. 26.2).

5. Select the type of analysis, the display format, the enzymes to be displayed, and which enzymes are to be used in the analysis.

6. Select "Analyze Sequence."

7. Analyze the results for useful restriction enzyme sites.

Please enter a title for this sequence:
Untitled sequence

Paste the DNA sequence into the box below

Please select the type of analysis you would like
◉ Linear sequence analysis
○ Circular sequence analysis
○ Find sites which may be introduced by silent mutagenesis

Please indicate how you would like the restriction sites displayed
☑ Map of restriction sites
☑ Table of sites, sorted alphabetically by enzyme name
☐ Table of sites, sorted sequentially by base pair number

Please indicate which enzymes to include in the display
◉ All enzymes
○ Enzymes not cutting
○ Enzymes cutting once
○ Enzymes cutting exactly ☐ times
○ Enzymes cutting at least ☐ times, and at most ☐ times
☑ Rainbow ▾ highlights for enzymes from the Standard ▾ polylinker

Please indicate which enzymes to include in the analysis
○ All enzymes in the database
◉ Only enzymes with recognition sites equal to or greater than 6 bases long
○ Only the following enzymes: AatI AatII Acc113I Acc16I Acc65I
Use the command, control, or shift key to select multiple entries
Analyze sequence | Clear sequence

Webcutter 2.0, copyright 1997 Max Heiman. Please send suggestions to maxwell@minerva.cis.yale.edu or visit the author's home page.

Figure 26.2 Webcutter restriction enzyme analysis page.

PROTOCOL 2 BLAST SEARCH FOR HOMOLOGOUS SEQUENCES

1. Locate GenBank at the NCBI website (http://ncbi.nlm.nih.gov).
2. Select "BLAST" at the top of the NCBI home page.
3. Select "Basic BLAST Search" under BLAST 2.0.
4. Choose program "blastn" and database "nr" (Fig. 26.3).
5. Choose the "FASTA" format.
6. Cut and paste the sequence from protocol 1 into the sequence box.
7. Select "Search."
8. Click "Format Results."
9. Observe the search results for identical and homologous sequences.

Websites are continually updated; if the NCBI home page is different from that illustrated, follow the new directions.

blastn is one of six programs in the BLAST family.

blastn compares the nucleotide sequence submitted with a nucleotide database.

blastn is designed for speed but automatically checks the complementary strand of the sequence queried.

nr selects the nonredundant databases at GenBank, EMBL, DDBJ, and Protein Data Bank (PDB).

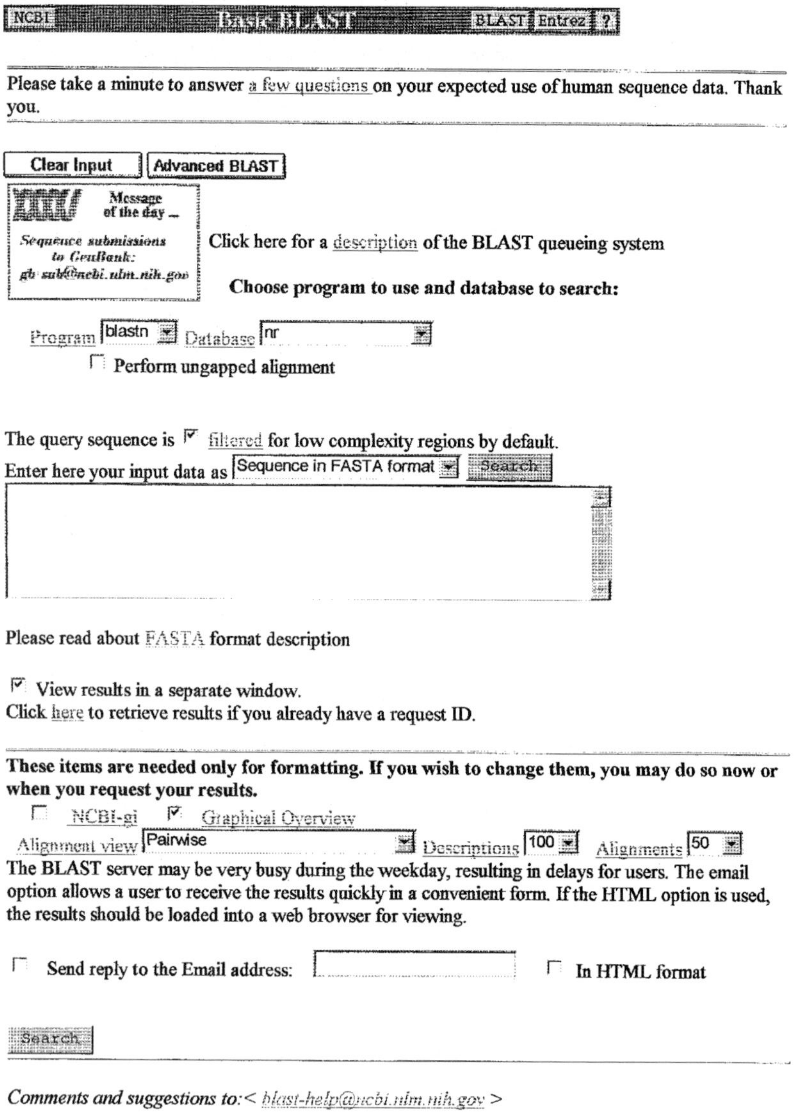

NCBI Basic BLAST BLAST Entrez ?

Please take a minute to answer a few questions on your expected use of human sequence data. Thank you.

Clear Input | Advanced BLAST

Message of the day ...
Sequence submissions to GenBank: gb-sub@ncbi.nlm.nih.gov

Click here for a description of the BLAST queueing system

Choose program to use and database to search:

Program blastn Database nr

Perform ungapped alignment

The query sequence is filtered for low complexity regions by default.

Enter here your input data as Sequence in FASTA format Search

Please read about FASTA format description

View results in a separate window.

Click here to retrieve results if you already have a request ID.

These items are needed only for formatting. If you wish to change them, you may do so now or when you request your results.

NCBI-gi Graphical Overview

Alignment view Pairwise Descriptions 100 Alignments 50

The BLAST server may be very busy during the weekday, resulting in delays for users. The email option allows a user to receive the results quickly in a convenient form. If the HTML option is used, the results should be loaded into a web browser for viewing.

Send reply to the Email address: In HTML format

Search

Comments and suggestions to: < blast-help@ncbi.nlm.nih.gov >

Figure 26.3 NCBI BLAST search Web page.

PROTOCOL 3 GENERATION OF AMINO ACID SEQUENCE FROM NUCLEOTIDE SEQUENCE INFORMATION

ExPASy is dedicated to the analysis of protein sequences and structures.

Websites are continually updated; if the ExPASy home page is different from that illustrated, follow the new directions.

1. Locate the ExPASy server (http://www.expasy.ch/).
2. Under Proteomics Tools, select "DNA → Protein."
3. Select "Translate."
4. Scroll down to the sequence box. Enter the sequence from step 1 by cutting and pasting it into the box (Fig. 26.4).

Verbose will yield an output in the standard style of a single letter for each amino acid but will spell out Met and stop codons wherever they occur in the sequence.

5. Set output format to "Verbose" ("Met," "Stop," space between residues).
6. Select "Translate sequence."
7. Observe all six sequences returned by the search for potential Met initiation sites and stop codons.

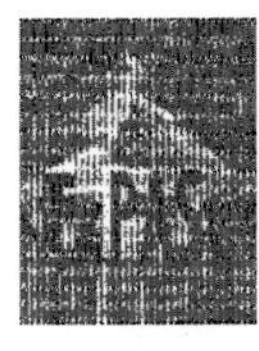

Translate tool

Translate is a tool which allows the translation of a nucleotide (DNA/RNA) sequence to a protein sequence.

Please enter a DNA or RNA sequence in the box below (numbers and blanks are ignored).

Output format: Verbose ("Met", "Stop", spaces between residues)

Reset or TRANSLATE SEQUENCE

Last modified 8/Feb/1999 by ELG

ExPASy Home page | Site Map | Search ExPASy | Contact us | Proteomics tools

Mirror sites: Australia Canada Taiwan

Figure 26.4 ExPASy translation analysis Web page.

References and General Reading

Altschul, S. F., W. Gish, W. Miller, E. W. Myers, and D. J. Lipman. 1990. Basic local alignment search tool. *J. Mol. Biol.* **215:**403–410.

Appel, R. D., A. Barroch, and D. F. Hochstrasser. 1994. A new generation of information retrieval tools: the example of the ExPASy WWW server. *Trends Biochem. Sci.* **19:**258–260.

Baxevanis, A. D., M. S. Boguski, and B. F. F. Ouellette. 1996. Computational analysis of DNA and protein sequences. *In* B. Birren, E. D. Green, S. Klapholz, R. M. Myers, and J. Roskams (ed.), *Genome Analysis: A Laboratory Manual,* vol. 1. *Analyzing DNA.* Cold Spring Harbor Laboratory Press, Cold Spring Harbor, N.Y. (http://clio1.cshl.org:80/books/chapters/analdna/ga1_chap.htm).

Brawley, S. H. 1999. Submission and retrieval of an aligned set of nucleic acid sequences. *J. Phycol.* **35:**433–437.

Bult, C. J. 1998. How to make sense of sequences. *Science* **282:**635–636.

Internet Sites

http://www.expasy.ch/ (ExPASy Molecular Biology Server, Swiss Institute of Bioinformatics)
http://www.firstmarket.com/cutter (Webcutter)
http://www.ncbi.nlm.nih.gov/ (National Center for Biotechnology Information)

Discussion

1. Describe, in detail, reasons for generating a restriction enzyme map of a DNA sequence.
2. Observe the restriction enzyme map. Did you identify any enzymes that do not cut the sequence or cut only one time?
3. Describe an experiment that would use the DNA sequence from this exercise and require a restriction enzyme that does not cut the sequence.
4. Describe an experiment that would use the DNA sequence from this exercise and require a restriction enzyme that cuts the sequence only one time.
5. How is a BLAST search done, and what is its purpose?
6. Observe the results of your BLAST search. Can you identify your sequence?
7. Are there any sequences in the same species that are homologous? What are they? Describe how this may occur.
8. Are there any sequences that are homologous but belong to a different species? What are they? Describe how this may occur.
9. What information is gained by translating a DNA sequence into its amino acid sequence?
10. Observe the results of the translation analysis. Can you identify the correct reading frame? Can you identify an initiation Met?

EXERCISE 27

Oligonucleotide Design by Computer

Key Concepts

1. Oligonucleotides are short synthetic fragments of DNA.
2. Oligonucleotides are important components of a polymerase chain reaction (PCR) because they identify the DNA fragment to be amplified.
3. Oligonucleotide length, base composition, and nucleotide sequence affect annealing temperature and function.
4. The melting temperature (T_m) of an oligonucleotide can be determined by the formula $T_m = [4(G+C) + 2(A+T)]$ °C, or the formula $T_m = 69.3°C + 0.41[\%(G+C)] - 650/l$, where l = length of the primer.

Goals

By the end of this exercise, you will be able to

- calculate the T_m of an oligonucleotide
- describe factors that affect oligonucleotide binding and function
- manually design primers for PCR
- design primers with computer analysis assistance available on the World Wide Web

Introduction

Oligonucleotides are short (usually 17 to 40 nucleotides), chemically synthesized sequences of single-stranded DNA. Usually, an investigator will design the oligonucleotide desired for a particular use, then contract to have it synthesized by one of the many biotechnology companies specializing in custom DNA synthesis.

Oligonucleotides are required for several different types of investigation. If a researcher wishes to screen a complementary DNA (cDNA) library for a very rare gene product, that is, a protein expressed at a very low level, it may be impossible to isolate the corresponding mRNA in order to make a cDNA probe. The investigator could isolate the protein first (see exercises 18 and 19), use proteases to digest it into peptides, and then determine the amino acid sequence of a few of the peptides. One only needs to determine a six- or seven-amino-acid sequence of the protein. Using the genetic code for the short peptide sequence, a researcher can design a synthetic oligonucleotide that would encode it; this oligonucleotide, once appropriately labeled for detection, can be used as a hybridization probe to locate the library clone containing the gene of interest. Of course, the degeneracy of the genetic code must be taken into account; in reality, a mixture of oligonucleotides must be synthesized to represent every possible nucleotide sequence coding for the amino acid sequence. A 20-mer oligonucleotide, representing six to seven amino acids, is sufficiently long to screen a cDNA library, so the task of designing suitable oligonucleotides is not as daunting as it appears.

For diagnosis of hereditary diseases, allele-specific oligonucleotides (ASO) may be designed, if both the normal and disease sequences are known. The ASO may be used as probes for Southern blot of genomic DNA or may be designed to amplify normal or disease loci, provided sufficient stringency is incorporated into the reaction. More stringent conditions are usually achieved by raising the reaction temperature.

Custom-designed oligonucleotides are used for making synthetic genes. If the desired synthetic gene is only 60 to 80 base pairs, then two complementary single-stranded DNA oligonucleotides may be synthesized, then annealed. Should the synthetic gene length exceed 300 base pairs, then individual oligonucleotides (20- to 60-mer each) are synthesized to incorporate overlapping complementary ends. When mixed, they will anneal to form a stable duplex; DNA ligase is used to join the ends, and the synthetic gene is formed.

Researchers may employ in vitro mutagenesis to study gene function. Oligonucleotides are used to direct site-specific mutagenesis. In one method, the wild-type gene of interest is cloned into a suitable plasmid vector, whose strands are then separated. A mutagenic oligonucleotide, designed so that the mutation desired is flanked by 10 to 15 wild-type bases, is annealed to the wild-type gene template within the vector. If the reaction condition stringency is sufficiently low, annealing will take place. DNA polymerase then extends the mutant strand, DNA ligase seals the nicks, and a heteroduplex plasmid is formed (one wild-type DNA strand, one mutant strand). A suitable *Escherichia coli* host is transformed with this vector; some transformed clones will carry the mutation.

Gene cassettes, which are synthetic double-stranded DNA oligonucleotides, each carrying a specific mutation, can be inserted and removed from a gene of interest to provide another method of in vitro mutagenesis. However, two restriction enzyme sites must be close together so that the cassette can be pasted in easily. Restriction enzyme sites can be incorporated into oligonucleotide design if they do not naturally occur where the investigator wishes to place a gene cassette.

The use of oligonucleotides as therapeutic agents is under investigation, and several promising lines of research have emerged. Antisense oligonucleotides of 15 to 20 nucleotides have been designed to bind to the specific messenger RNAs (mRNAs) of genes implicated in cancer and several other diseases. When the oligonucleotides, modified to resist degradation in the cell, bind to the corresponding mRNA, RNase H breaks down the mRNA portion of the hybrid molecule (heterodimer). Consequently, translation of the protein involved in the disease is curtailed or eliminated. Liposomes are used to enclose and deliver the synthetic oligonucleotide to cells; tagging the liposome with appropriate cell-receptor binding proteins can target specific tissues. Similar antisense technology has been used to abbreviate viral infections.

DNA sequencing and PCR are, perhaps, the applications most commonly employing oligonucleotides (see exercises 10 and 28). PCR has become a widely used technique in many research and clinical laboratories. It can be used to detect polymorphisms (exercise 10), for DNA cloning (exercise 14), to study homologous genes (exercise 23), for DNA sequencing (exercise 28), as well as for diagnosis of infectious diseases. Since its development in the 1980s, PCR has truly revolutionized many fields of science. For a brief review, see exercise 10.

PCR depends on a specific pair of oligonucleotide primers to flank the target DNA sequence to be amplified. Design of oligonucleotides, and corresponding adjustment of the annealing temperature, allows a researcher to amplify target DNA whose bracketing sequences may not match the primer perfectly.

Guidelines for Efficient Primer Sequences

General guidelines are used to choose primer sequences.

- Primers are usually 17 to 28 bases in length.
- The base composition should be about 50 to 60% (G+C).

- The base sequence of a primer should be random, without unusual stretches of one base repeated.
- The 3′ end of a primer should be G, C, GC, or CG to increase annealing efficiency.
- Avoid three or more runs of G and C nucleotides at the 3′ end of the primer to avoid stable annealing of G-C sequences and mispriming of DNA.
- T_ms between 55 and 80°C are best.
- Primer pairs should not be complementary at the 3′ end to avoid primers pairing with each other (primer-dimers).
- Avoid self-complementary sequences within a primer that may form hairpin structures.

Melting Temperature and Primer Annealing

The T_m of an oligonucleotide may be determined using the formula: $T_m = [4(G+C) + 2(A+T)]$ °C, or the formula $T_m = 69.3°C + 0.41[\%(G+C)] - 650/l$, where l = length of the primer. An annealing temperature about 5°C below the T_m is used in the PCR. These determinations provide guidelines for annealing temperature; primer pairs used for the first time will require experimentation for determination of optimal annealing temperature.

As with any hybridization reaction, the ionic strength of the reaction buffer also plays a role in promoting DNA-DNA interactions. Using up to 50 mM KCl will facilitate primer annealing. Higher concentrations will inhibit *Taq* polymerase function.

Human Immunodeficiency Virus

Human immunodeficiency virus type 1 (HIV-1) is the causative agent of the AIDS (acquired immune deficiency syndrome) pandemic. Some estimates of global infection rates predicted that by 2000, over 40 million people would be infected with HIV-1. The mortality rate from AIDS approaches 100%. HIV-1 infection risk can be greatly lowered by condom use (however, only abstinence or mutual monogamy with an uninfected partner eliminates the risk of sexual transmission) and avoidance of intravenous drug abuse and shared needles and syringes.

HIV-1 is a lentivirus, a member of the retrovirus family. Retroviruses are RNA viruses possessing reverse transcriptase, so they synthesize a DNA copy of the viral genome. This DNA copy integrates into the human host cell genome. The lentiviruses have more complex genomes than other retroviruses; in addition to the usual structural genes, *gag, pol,* and *env,* HIV-1 possesses six additional regulatory and accessory genes: *tat, rev, nef, vif, vpr,* and *vpu.* The *gag* gene encodes four nucleocapsid proteins, translated as a single precursor, which HIV-1 protease cleaves. One of the proteins, p25, is the most prevalent component of the capsid. Regions of the *gag* gene are highly conserved in HIV-1 isolates, despite the rapid mutation rate found in the virus.

Screening for HIV-1 infection depends on the presence of HIV-1 antibodies, detected by ELISA (enzyme-linked immunosorbent assay) (see exercise 19). Antibodies appear in the serum several weeks to months after infection. However, when it is important to ascertain HIV-1 infection before antibodies are present or when the immune response is undetectable, PCR may be used to target the HIV-1 *gag* gene.

For this laboratory exercise, you will design a pair of PCR oligonucleotides that could be used to amplify the HIV-1 *gag* sequence.

Lab Language

annealing
antisense oligonucleotide
ASO
gene cassette
melting temperature
mispriming
oligonucleotide
primer
primer-dimer

Materials

Materials per pair of students
computer with Internet access
Web browser (such as Netscape or Internet Explorer)
floppy disks (optional)
printer
calculator

Protocols

PROTOCOL 1 OBTAINING THE HIV SEQUENCE TO BE AMPLIFIED

Websites are continually updated; if the NCBI home page is different from that described, follow the current directions.

1. Locate the National Center for Biotechnology Information (NCBI) (http://www.ncbi.nlm.nih.gov/).
2. Select "Search GenBank" for "K02007."

This accession number will give you the complete proviral genome of the HIV-1 isolate ARV-2/SF2.

3. Select "Go."
4. Under "Current Query" select "Retrieve [number] Documents."
5. Select "Display GenBank report" for K02007.
6. What information can be obtained when searching for a specific accession number?
7. Copy the nucleotide sequence from positions 1400 to 1800.

Once you have copied the sequence, you may quit the Web browser.

8. Open a word-processing file and paste the sequence into it. Save the document.
9. Select "double space" and print out the sequence.

This is the conserved gag *gene.*

10. Find and delineate (highlight or underline) the positions 1551 through 1665.

PROTOCOL 2 MANUAL DESIGN OF PCR OLIGONUCLEOTIDE PRIMER PAIR

1. Study the highlighted *gag* gene sequence and its neighboring codons.
2. Write in the complementary strand beneath it.

The forward primer goes 5′ to 3′ from the beginning of the selected region and binds to the complementary strand.

The reverse primer goes 3′ to 5′ from the end of the selected region and binds to the given strand.

A single-stranded DNA primer can fold back on itself to form a double-stranded DNA molecule that will probably not prime the PCR.

Complementarity between the forward and reverse primers will produce primer-dimers and failure to amplify the selected region.

3. Construct a forward primer and a reverse primer to amplify this sequence using the following criteria:
   - Primer length = 20 nucleotides.
   - T_m = 55 to 59°C.
   - Avoid hairpin loops.
   - Avoid long repeats of a single base.
   - Avoid complementarity between the forward and reverse primer, especially at the 3′ ends.
   - Aim for 50 to 60% G+C content.

Count the number of bases of each type; calculate the percent G+C.

4. Calculate the T_m using the following formulas:
   - $T_m = [4(G+C) + 2(A+T)]$ °C
   - $T_m = 69.3°C + 0.41[\%(G+C)] - 650/l$, where l = length of the primer.

5. Write your primers below.

   Forward primer:

   Reverse primer:

PROTOCOL 3 COMPUTER-ASSISTED OLIGONUCLEOTIDE PRIMER DESIGN

1. Open your word-processing file that contains the desired *gag* sequence (protocol 1).
2. Highlight, select, and copy positions 1400 to 1800.
3. Access your Web browser and go to "*Primers!* for the world wide web" (http://www.williamstone.com/primers/index.html) (Fig. 27.1).

   Several primer design programs are available to the public free of charge.

4. Select "*Primers!* for the world wide web (Primer Design)."

   Websites are continually updated; if different from that illustrated, follow the new directions.

5. Adjust the parameters on the left screen to reflect primer length and T_m used in protocol 2.
6. Enter your sequence in the box by copying and pasting from your word-processing document. Delete any numerals.

   Numerals will not be recognized as valid entries.

7. Under "4. Generate Primers," select "pick primers" (Fig. 27.2).
8. Examine and choose a primer pair as instructed on the results page. You may select or deselect any of the available analysis tools.
9. Copy your selected pair below.

   Forward primer:

   Reverse primer:

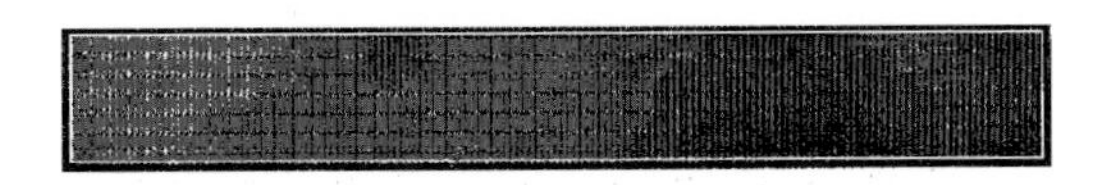

Welcome!

Welcome to *Primers!* for the world wide web. This site hosts a comprehensive PCR primer design application, which you may use as often as you like, free of charge. We are aggressively seeking oligo manufacturing companies to join us as partners, so that once you finish designing your primers, you may order them directly from the manufacturer by clicking a button on this site. If you would like to recommend a partner, please get in touch with us.

If you would like to analyze a primer which you've already designed, please use our Primer Calculator to do so.

To begin the primer **design** process, just follow the three steps below.

1. Adjust parameters

The parameters to the left are adjustable. Once you are satisfied with the values, you may save the settings as a cookie so that the next time you return, you will not have to make any adjustments. Click on the *Save Settings* icon to do so.

Some versions of Netscape have a bug which causes the settings to the left to appear blank the first time you visit this page. Click on the *Reset Settings* icon to fix this problem.

2. Enter your sequence

Paste or type a sequence into the box below. Legal sequences consist of a string of the characters `[A,C,G,T,N,a,c,g,t,n]`. White space is allowed (including spaces and carriage returns). Please note that this transmission is not secure.

3. Option - Restrict Generation

You may optionally decide to restrict primer generation to specific regions on your sequence. If you decide to do so, fill out the text fields below. All specifications should be 1-based, (i.e., the first base of your sequence is counted as number 1, not as number 0).

To generate primers freely, skip this section.

```
Only generate forward primers from [    ] (5') to [    ] (3').
Only generate reverse primers from [    ] (3') to [    ] (5').
```

4. Generate Primers

Click the button below to begin primer generation. This may take anywhere from 5 seconds upwards, depending on the size of your input sequence. A 300bp sequence averages 30 seconds of processing time.

Pick Primers

Figure 27.1 Primer design from "*Primers!* for the world wide web."

✓ Base Count
✓ Composition
✓ Colorize Product

3. Choose Primer Pair

Examine the two tables below to select one forward and one reverse primer for the analyses you specified in step 1, above.

Forward Primers

Click on the button to the left of any primer to make it the active forward primer.

Primer	Base Number	Tm°C
AGCAACCTCAAACAGACACC	11 .. 30	57.77
GCAACCTCAAACAGACACCA	12 .. 31	59.73
ACCTCAAACAGACACCATGG	15 .. 34	58.41
CCTCAAACAGACACCATGGT	16 .. 35	58.41
CTCAAACAGACACCATGGTG	17 .. 36	57.47
TCAAACAGACACCATGGTGC	18 .. 37	60.58
CAAACAGACACCATGGTGCA	19 .. 38	61.63
AAACAGACACCATGGTGCAC	20 .. 39	59.45
AACAGACACCATGGTGCACC	21 .. 40	61.90
ACAGACACCATGGTGCACCT	22 .. 41	61.50
AGACACCATGGTGCACCTGA	24 .. 43	62.65
GACACCATGGTGCACCTGAC	25 .. 44	62.53
ACACCATGGTGCACCTGACT	26 .. 45	61.50
CACCATGGTGCACCTGACTC	27 .. 46	62.65
CATGGTGCACCTGACTCCTG	30 .. 49	62.78

Reverse Primers

Click on the button to the left of any primer to make it the active reverse primer.

Primer	Base Number	Tm°C
AGGCCCTTCATAATATCCCC	561 .. 542	59.12
GGCCCTTCATAATATCCCCC	560 .. 541	61.54
GCCCTTCATAATATCCCCCA	559 .. 540	60.85
CCCTTCATAATATCCCCCAG	558 .. 539	58.20
TGGACTTAGGGAACAAAGGA	529 .. 510	57.24

4. Analyze Primers

Click the *Analyze* button, below, to begin the analysis. Alternatively, you may enter a different sequence by clicking the *Back* button.

[Back] [Analyze]

Figure 27.2 Results obtained from a limited primer design scheme.

PROTOCOL 4

COMPARISON OF MANUAL AND COMPUTER-ASSISTED OLIGONUCLEOTIDE PRIMER DESIGN

1. Write the primer pairs you obtained manually.

   Forward primer:

   Reverse primer:

2. Write the primer pairs obtained by using computer assistance.

   Forward primer:

   Reverse primer:

3. List similarities and differences.

4. Access "*Primers!* for the world wide web."

5. Select "Primer calculator."

6. Enter the primers you designed manually.

7. Select "Evaluate" (Fig. 27.3).

Figure 27.3 Primer analysis obtained with D1S80 locus primers.

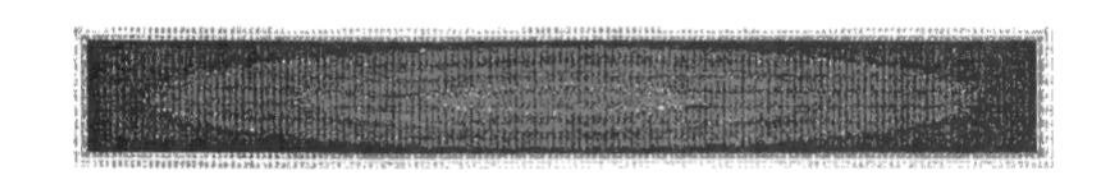

Forward Primer		Reverse Primer	
GAAACTGGCCTCCAAACACTGCCCGCCG		CGTTCCCCGTGCACGTAGAGGTTGTTCTG	
A - 7 (25.00%)	C - 12 (42.86%)	A - 3 (10.34%)	C - 8 (27.59%)
G - 6 (21.43%)	T - 3 (10.71%)	G - 9 (31.03%)	T - 9 (31.03%)
N - 0 (0.00%)		N - 0 (0.00%)	
Length: - 28 bases	Mlclr Weight - 8484 daltons (g/M)	Length: - 29 bases	Mlclr Weight - 8884 daltons (g/M)
T_m		T_m	
Nearest Neighbor T_m (standard):	80.44	Nearest Neighbor T_m (standard):	76.16
%GC Method T_m	67.24	%GC Method T_m	65.75
2(A+T) + 4(G+C) T_m:	92.00	2(A+T) + 4(G+C) T_m:	92.00

Forward Primer **Reverse Primer**

Primers may fold and anneal against themselves. A likely folding scenario is displayed below.

```
5' GAAACTGGCCTCCA 3'                 5' --CGTTCCCCGTGCACG 3'
        | |     ||                                |  |  |  |
3' ----GCCG-CCCGTCACAA 5'            3' GTCT-TGTTGGAGAT 5'
```

Primers may anneal with other copies of themselves. Below is a candidate alignment.

```
5' GAAACTGGCCTCCAAACACTGCCCGCCG 3'      5' --------------CGTTCCCCGTGCACGTAGAGGT
       | |       ||        ||    |||                    ||| |  |  |  |  |||
3' ----GCCG-CCCGTCACAAACCTCCGGTCAAAG 5' 3' GTCTTGTTGGAGATGCACGTGCCCCTTGC------
```

Forward primers may anneal with with reverse primers. A likely alignment is shown below.

```
5' -----------GAAACTGGCCTCCAAACACTGCCCGCCG 3'
              |     |  ||     ||||    |
3' CGTTCCCCGTGCACGTAGAGG----TTGTTCTG 5'
```

I would like to order these primers from my supplier

References and General Reading

Glick, B. R., and J. J. Pasternak. 1998. *Molecular Biotechnology: Principles and Applications of Recombinant DNA,* 2nd ed. ASM Press, Washington, D.C.

Innis, M. A., D. H. Gelfand, J. J. Sninsky, and T. J. White. 1990. *PCR Protocols: a Guide to Methods and Applications.* Academic Press, New York, N.Y.

Oh, C.-Y., S. Kwok, S. W. Mitchell, D. H. Mack, J. J. Sninsky, J. W. Krebs, P. Feorino, D. Warfield, and G. Schochetman. 1988. DNA amplification for direct detection of HIV-1 in DNA of peripheral blood mononuclear cells. *Science* **238:**295–297.

Prescott, L. M., J. P. Harley, and D. A. Klein. 1996. *Microbiology,* 3rd ed. William C. Brown Publishers, Dubuque, Iowa.

Turner, P. C., A. G. McLennan, A. D. Bates, and M. R. H. White. 1997. *Instant Notes in Molecular Biology.* BIOS Scientific Publishers, Springer-Verlag, New York, N.Y.

Watson, J. D., M. Gilman, J. Witkowski, and M. Zoller. 1992. *Recombinant DNA,* 2nd ed. W. H. Freeman & Co., New York, N.Y.

Internet Site
http://www.williamstone.com/primers/index.html (Williamstone Enterprises, *Primers!* for the World Wide Web)

Discussion

1. Discuss some of the applications for which oligonucleotides are necessary.
2. What are three problems encountered with oligonucleotide design, and how are these problems solved?
3. Evaluate your choice of primers, designed manually, in protocol 2.
4. Discuss the pros and cons of using computers for oligonucleotide design.
5. Why are primers to a conserved region of the *gag* gene used in diagnostics?

SECTION IX

Advanced Techniques

EXERCISE 28

DNA Sequencing

Key Concepts

1. DNA sequence analysis is important for understanding gene structure, regulation, and function.
2. Two methods for determining DNA sequence have been developed: the dideoxy chain termination method and the chemical cleavage method.
3. The entire genome sequences of such organisms as the eukaryotes *Caenorhabditis elegans* (a roundworm) and *Saccharomyces cerevisiae* (a yeast), the prokaryotes *Escherichia coli, Chlamydia trachomatis, Helicobacter pylori,* and *Rickettsia prowazekii,* and many viruses have been determined.
4. The goal of the Human Genome Project is to sequence the entire human genome by 2003.
5. The race to sequence the human genome has led to technological breakthroughs in DNA sequencing technology.

Goals

By the end of this exercise, you will be able to

- understand the basic differences between the dideoxy method and the chemical cleavage method of DNA sequencing
- discuss how polymerase chain reaction (PCR) can be used in DNA sequencing
- read a dideoxy DNA sequencing gel

Introduction

DNA Sequencing

The nucleotide sequence information of a particular DNA molecule is key to understanding gene structure and function. Knowledge of DNA sequence can provide the researcher with a mechanism for generating restriction enzyme maps, which then allow for manipulation and further study of a gene.

There are two methods for determining DNA sequence. In 1980, Walter Gilbert and Frederick Sanger shared the Nobel Prize in chemistry for developing these DNA sequence methodologies. Sanger, along with S. Nicklen and A. Coulson, developed the enzymatic dideoxy sequencing method. Independently, A. Maxam and W. Gilbert developed the chemical cleavage method for DNA sequencing.

Figure 28.1 (A) A dideoxynucleotide lacks a 3′-hydroxyl group. (B) Nucleoside triphosphates contain a 3′-hydroxyl. (Reprinted from B. R. Glick and J. J. Pasternak, *Molecular Biotechnology: Principles and Applications of Recombinant DNA,* 2nd ed. ASM Press, Washington, D.C., 1998, with permission.)

Dideoxy DNA Sequencing

The dideoxy or chain termination method of DNA sequencing uses 2′,3′-dideoxynucleoside triphosphates (ddNTPs). When a ddNTP is incorporated into the growing DNA strand, chain elongation is terminated because the ddNTP lacks a 3′-hydroxyl group (Fig. 28.1 and 28.2) to which the new nucleotide must bond to continue synthesizing a longer DNA molecule. Typically, the

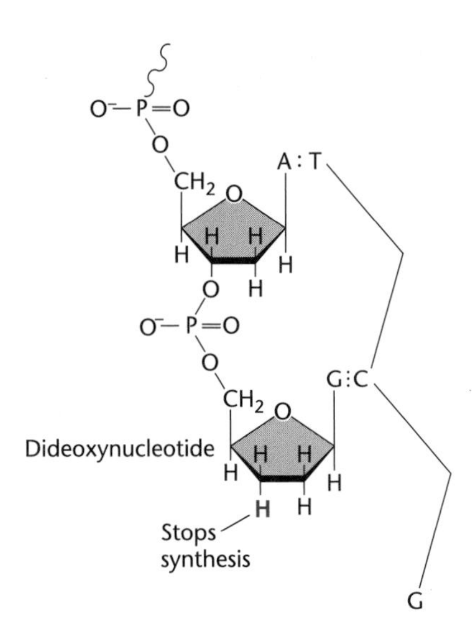

Figure 28.2 Dideoxynucleotides terminate DNA synthesis because a free 3′-hydroxyl group is not available to chemically bond to additional nucleoside triphosphates. (Reprinted from B. R. Glick and J. J. Pasternak, *Molecular Biotechnology: Principles and Applications of Recombinant DNA,* 2nd ed. ASM Press, Washington, D.C., 1998, with permission.)

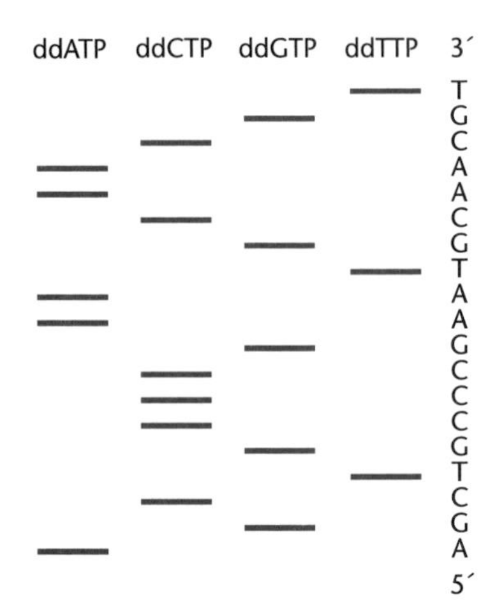

Figure 28.3 Schematic diagram of a silver-stained dideoxy sequencing gel. Four reactions were carried out, each with a different chain terminator (ddATP, ddCTP, ddGTP, or ddTTP). Bands were resolved by acrylamide gel electrophoresis and stained. Sequence is read from the "bottom" (short fragments) up. (Reprinted from B. R. Glick and J. J. Pasternak, *Molecular Biotechnology: Principles and Applications of Recombinant DNA,* 2nd ed. ASM Press, Washington, D.C., 1998, with permission.)

DNA primer (used to initiate synthesis) is radiolabeled. Once the DNA sequencing reaction has been carried out, the terminated chains of newly synthesized DNA are resolved via polyacrylamide gel electrophoresis. This type of gel can resolve DNA fragments that differ from one another by only one nucleotide in length. By using four different sequencing reactions, each with a different chain terminator—ddATP, ddCTP, ddGTP, and ddTTP—the entire sequence of a DNA fragment can be determined (Fig. 28.3).

Fluorescent Dideoxy DNA Sequencing

The method of dideoxy DNA sequencing has been improved by the use of fluorescent ddNTPs. Each terminator molecule is labeled with a different fluorescent dye. In this manner, all four ddNTPs and all four dNTPs (deoxynucleoside triphosphates) are used together in one sequencing reaction. To resolve the DNA sequence, the reaction products are separated via polyacrylamide gel electrophoresis, but as the different length fragments are separating, they electrophorese past a laser and detection system. The laser excites the four different fluorescently labeled molecules. The detector records the color of fluorescence that indicates the ddNTP incorporated into each fragment. This has eliminated the need for radioactive materials and has added automation to DNA sequencing.

PCR Applied to Dideoxy DNA Sequencing

Since dideoxy DNA sequencing relies on the enzyme DNA polymerase, it was only logical that PCR would be used to perform DNA sequencing. Heat-stable DNA polymerases allow the use of smaller amounts of DNA template. Repeated annealing, denaturation, and elongation in the presence of excess primer, ddNTP, and dNTP provide for repeated use of the template DNA. PCR may be used for sequencing with both radiolabeled primers and fluorescently labeled ddNTP. Sequencing products are resolved via polyacrylamide gel electrophoresis as in the original method.

The Chemical Cleavage Method of DNA Sequencing

The Maxam and Gilbert chemical cleavage method of DNA sequencing utilizes chemical differences in the purines (adenine and guanine) and the pyrimidines (cytosine and thymidine). Formic acid, hydrazine, or dimethyl sulfate (DMS) is used to modify specific bases, and then piperidine is used to break the strands at the modified bases. The resulting DNA fragments are resolved via polyacrylamide gel electrophoresis.

Formic acid protonates the nitrogens in the purine rings of adenine (A) and guanine (G). Addition of piperidine cleaves DNA at both A and G residues. DMS methylates nitrogen 7 of G, which allows piperidine to break the DNA strand at G residues only. Comparing the DMS-treated DNA sequencing lane with the formic acid-treated DNA sequencing lane allows one to determine whether the sequence break is at an A or a G residue. Hydrazine treatment of DNA provides a mechanism of cleaving DNA at the pyrimidine residues. When DNA is treated with hydrazine in the presence of sodium chloride, only the cytosine (C) residue is modified and subject to cleavage by piperidine. In the absence of sodium chloride, both the C and thymidine (T) moieties are modified and cleaved with piperidine. Comparing the resulting fragments generated in the presence or absence of sodium chloride allows one to determine whether the break is at a C or a T residue.

The Human Genome Project

Begun in the 1980s, the Human Genome Project had the goal of sequencing the entire human genome by 2005. Since the human genome is estimated to be 3 billion base pairs in length, determining the exact nucleotide sequence is a monumental task involving many researchers in laboratories worldwide. This project is only partially completed, but already researchers have set an earlier deadline of

2003 for determination of the entire sequence. This will require tremendous improvements in sequencing technology as well as reduction of cost.

Some of the first organisms' genomes to be sequenced were viruses and bacteria. Their genomes are generally much smaller than eukaryotic ones. These include *E. coli* (genome size, 3 million bases), *C. trachomatis, H. pylori* (associated with ulcers), and *R. prowazekii* (typhus). The first eukaryote to be sequenced was the yeast *S. cerevisiae* (genome size, 12.5 million bases) in 1996. The only other eukaryote, and the first multicellular organism, to have its genome sequenced was the nematode (roundworm) *C. elegans* (genome size, 97 million bases). This sequencing project was essentially finished in late 1998. Plant genome sequences are also being pursued. The sequence of *Arabidopsis thaliana* was finished in 1999, while maize (corn) and wheat, with genome sizes larger than that of humans, will have to wait.

The Future of DNA Sequencing Technology

One goal of the Human Genome Project, necessary to complete the sequence by 2003, is the development of increasingly efficient methodologies for sequencing. This includes automated fluorescent sequencers. But technology must evolve even further, to fully automating sequencing technologies using tiny biochips. In the future, individuals may carry their own DNA sequence with them on a portable biochip the size of a postage stamp, which will enable rapid diagnosis and treatment of disease on an individual basis.

Lab Language

biochip
chain terminator
chemical cleavage
dideoxynucleotides
DNA polymerase
Human Genome Project
fluorescent ddNTP
radiolabeled primer

Laboratory Safety

Review the sections on physical, biological, and chemical hazards in "Laboratory Safety." Special considerations for this exercise:

gel electrophoresis
acrylamide-bisacrylamide
N,N,N',N'-tetramethylethylenediamine (TEMED), ammonium persulfate (APS)

Laboratory Sequence

Day 1: PCR sequencing (protocol 1)
Day 2: Gel electrophoresis (protocols 2 and 3)
Silver staining (protocol 4) (fixing step may incubate overnight)
Drying of gel
Day 3: Analysis of results

Materials

DNA Sequencing Reaction

Class equipment and supplies
thermal cycler
sequencing kit (Promega)

Students will need to prepare
sequencing reaction mixture

Materials per pair of students
1 aliquot of pGEM DNA for sequencing
1 aliquot of 5× DNA sequencing buffer
1 aliquot of forward primer
1 aliquot of dH_2O
1 aliquot of *Taq* DNA polymerase

1 aliquot of dNTPs/ddATP
1 aliquot of dNTPs/ddCTP
1 aliquot of dNTPs/ddGTP
1 aliquot of dNTPs/ddTTP
gloves
PCR microcentrifuge tubes
0.5- to 10-µl micropipette and sterile tips

DNA Sequencing Gel Electrophoresis

Class equipment and supplies
8% acrylamide-bisacrylamide, 19:1
10% APS
TEMED
10× Tris-borate-EDTA (TBE)
Pasteur pipettes and bulbs
microcentrifuge
heat block or boiling water bath

Students will need to prepare
300 ml of 1× TBE

Materials per pair of students
1 aliquot of sequencing loading buffer
vertical gel electrophoresis system with glass plates, spacers, comb, chamber
gloves
microcentrifuge tubes
0.5- to 10-µl micropipette and sterile tips
100- to 1,000-µl micropipette and sterile tips
power supply capable of 250 volts (shared by 2 to 4 groups)
50-ml graduated cylinder
500-ml graduated cylinder

Protocols

PROTOCOL 1 PCR SEQUENCING

You will prepare four different sequencing reaction mixtures, each with a different ddNTP chain terminator.

Standard primers are available to sequences flanking cloning sites in plasmid vectors.

1. Prepare the sequencing mixture and template.

pGEM DNA, 1 µg/µl	4.0 µl
5× DNA sequencing buffer	5.0 µl
pUC/M13 forward primer, 4.5 pmol	3.6 µl
dH_2O	3.4 µl
Taq DNA polymerase, 5 U/µl	1.0 µl

2. Mix and pop spin.
3. Prepare the sequencing reaction tubes by labeling four tubes with A, C, G, and T.

Each mixture must contain all four dNTPs, but only one ddNTP.

4. Pipette 2 µl of the appropriate d/ddNTP mix into each tube.
5. Add 4 µl of the sequencing mixture and template from step 1 to each of the four reaction tubes.

Some thermal cyclers are equipped with a heated lid.

6. Pop spin the four sequencing reaction tubes and then add a drop of mineral oil to each tube.

Cycle 1 contains an extended 95°C step to fully denature template DNA.

7. Place the tubes in the thermal cycler. Conditions for cycling are:

cycle 1	step 1	95°C for 2 minutes
	step 2	42°C for 30 seconds
	step 3	70°C for 1 minute
cycles 2 to 30	step 1	95°C for 30 seconds
	step 2	42°C for 30 seconds
	step 3	70°C for 1 minute

Hold at room temperature or 4°C.

PROTOCOL 2 PREPARATION OF THE SEQUENCING GEL

Once the thermal cycler has completed the sequencing reactions, the DNA fragments are resolved on a polyacrylamide gel.

1. **Wearing gloves,** clean the glass plates with soap and water. Rinse well.

Clean plates help prevent bubble formation when pouring acrylamide gels.

2. Assemble the gel apparatus as shown in Fig. 28.4. The short glass plate should face outward and the longer plate inward. The clamp knobs should face inward.

Align the glass plates and spacers on a flat surface.

3. Pipette water between the glass plates to check that the apparatus has been assembled correctly and is leak proof. Pour the water out and blot with a paper towel.

The glass plates and spacers should form a watertight seal against the gasket.

4. **Wearing gloves,** mix the reagents together for 8 ml of an 8% polyacrylamide gel in a disposable tube.

Acrylamide is a neurotoxin: ***avoid skin contact.***

8% acrylamide-bisacrylamide	8.0 ml
10% APS	80 μl
TEMED	8 μl

APS initiates polymerization.

TEMED speeds up the polymerization process.

5. Immediately "pour" the gel using a Pasteur pipette.

6. Make sure there are no bubbles in the gel. Sharply tap the glass to force any bubbles to the top of the gel.

7. Insert the comb and make sure that there are no bubbles trapped beneath the teeth.

8. Save your tube containing the acrylamide solution. When the solution in the tube has solidified, the gel has solidified.

Begin preparation of your samples while the gel is solidifying.

9. Once the gel has solidified, transfer your gel to the electrophoresis chamber. Two groups will share an electrophoresis chamber to electrophorese two gels simultaneously.

10. Prepare 300 ml of 1× TBE from a 10× stock solution. One gel chamber requires 300 ml.

11. Add 1× TBE to the inside chamber so it covers the wells, and pour running buffer into the bottom of the chamber until it makes contact with the bottom of the gel.

12. *Carefully* remove the comb and rinse the wells with 1× TBE.

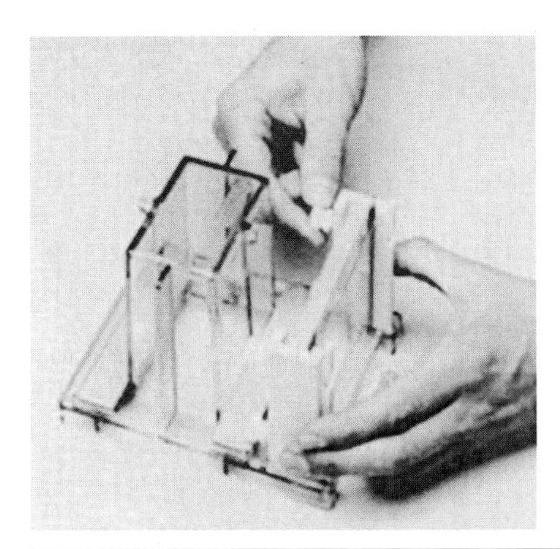

Figure 28.4 Assembly of gel apparatus for the DNA acrylamide gel. (Reprinted from Bio-Rad Laboratories, Hercules, Calif., with permission.)

PROTOCOL 3 Polyacrylamide Gel Electrophoresis

Oil will interfere with gel loading.

1. Carefully transfer 5 to 6 μl of each sequencing reaction mixture into clean tubes labeled A, C, G, and T.
2. Add 3 μl of sequencing loading buffer to each tube. Pop spin to mix.

Heating denatures DNA strands.

3. Heat the reactions at 70°C for 2 minutes, *just before loading the sequencing gel.*

Save extra reaction mixtures at –20°C for additional analysis.

4. Load 2 μl of each sequencing reaction mixture into separate wells.
5. Electrophorese the gel at 200 volts for approximately 45 minutes.

PROTOCOL 4 SILVER STAINING OF THE SEQUENCING GEL

1. After electrophoresis, carefully remove the sequencing gel from the electrophoresis chamber.
2. Remove the spacers and separate the glass plates.

The gel may be stored overnight in fix solution.

3. Carefully place the gel in fix solution for 20 minutes.

Longer rinses will result in faint band signals.

4. Rinse the gel with water three times, for 2 minutes each time, while gently agitating. Allow the gel to drain briefly between each rinse.
5. Place the gel in staining solution and gently shake for 30 minutes.
6. Rinse the gel *for no longer than 5 seconds* in water.
7. Immediately place the gel in chilled developing solution. Gently shake until the DNA bands become visible.
8. Transfer the gel to fresh developing solution for an additional 2 to 3 minutes.
9. Place the gel in fix solution for 2 to 3 minutes with gentle shaking.
10. Rinse the gel with water two times, for 2 minutes each time.
11. Dry the gel between sheets of gel-drying film supported by appropriate frames. See protocol 9 in exercise 18 or protocol 6 in exercise 19.
12. Read the sequencing gel from the bottom to the top (shorter to longer) by noting which reaction lanes, in order, have bands (Fig. 28.3).

References and General Reading

Ausubel, F. M., R. Brent, R. E. Kingston, D. D. Moore, J. G. Seidman, J. A. Smith, and K. Struhl (ed.). 1994. *Current Protocols in Molecular Biology.* Wiley Interscience, New York, N.Y.

The *C. elegans* Sequencing Consortium. 1998. Genome sequence of the nematode *C. elegans*: a platform for investigating biology. *Science* **282:**2012–2018.

Collins, F. S., A. Patrinos, E. Jordan, A. Chakravarti, R. Gesteland, L. Walters, and members of the DOE and NIH planning groups. 1998. New goals for the U.S. Human Genome Project 1998–2003. *Science* **282:**682–689.

Gale, M. D., and K. M. Devos. 1998. Plant comparative genetics after 10 years. *Science* **282:**656–659.

Maxam, A. M., and W. Gilbert. 1977. A new method for sequencing DNA. *Proc. Natl. Acad. Sci. USA* **74:**560–564.

Meinke, D. W., J. M. Cherry, C. Dean, S. D. Rounsley, and M. Koornneef. 1998. *Arabidopsis thaliana*: a model plant for genome analysis. *Science* **282:**662, 679–682.

Sanger, F., S. Nicklen, and A. R. Coulson. 1977. DNA sequencing with chain-terminating inhibitors. *Proc. Natl. Acad. Sci. USA* **74:**5463–5467.

Stephens, R. S., S. Kalman, C. Lammel, J. Fan, R. Marathe, L. Aravind, W. Mitchell, L. Olinger, R. L. Tatusov, Q. Zhao, E. V. Koonin, and R. W. Davis. 1998. Genome sequence of an obligate intracellular pathogen of humans: *Chlamydia trachomatis*. *Science* **282:**754–759.

Service, R. F. 1998. Coming soon: the pocket DNA sequencer. *Science* **282:**399–401.

Discussion

1. List five similarities between dideoxy and chemical cleavage DNA sequencing.
2. List five differences between dideoxy and chemical cleavage DNA sequencing.
3. Describe how dideoxynucleotides are used in DNA sequencing reactions.
4. How does your sequencing result compare with the expected DNA sequence? Discuss your answer.
5. How does incorporation of a ddNTP correspond to the DNA sequencing template?
6. Why do standard primers flank multiple cloning sites in plasmid vectors?

EXERCISE 29

DNA Library Construction

Key Concepts

1. A DNA library contains all the genetic material of a single species, broken into small fragments and cloned into a recombinant vector.
2. DNA libraries may be constructed using genomic DNA (gDNA) or DNA complementary to RNA (cDNA), and therefore lacking intron sequences.
3. DNA libraries may be used to isolate specific gene sequences of interest or may be used to sequence DNA fragments whose function is not yet known.
4. Screening DNA libraries for a specific insert may be performed by DNA hybridization or by examining clones for protein expression.

Goals

By the end of this exercise, you will be able to

- describe how gDNA and cDNA libraries are made
- discuss how DNA libraries may be used
- explain how a gene is selected from a DNA library using filter hybridization
- prepare a bacterial gDNA library
- understand how an expression library is screened

Introduction

DNA Libraries

A DNA library is a collection of yeast or bacterial cells, each containing recombinant vectors that carry DNA inserts from a single species. Depending on the genome size of the species studied, a library might require thousands of to a few million clones. DNA is isolated from the organism of interest, cut into workable fragments by restriction enzymes, cloned into a suitable vector, reproduced, and introduced into a host cell. The host replicates the DNA of interest and may express proteins. The library can be screened to locate the specific clone bearing the gene of interest.

The library, however, may be constructed with gDNA or with cDNA, which is DNA copied from an organism's messenger RNA (mRNA). A gDNA library contains all the DNA sequences of an organism, including regulatory sequences, exons, and introns, if the organism is eukaryotic. A cDNA library consists only of gene coding sequences, as represented by mRNA from a particular cell type. A distinct advantage of cDNA libraries is that the introns present in eukaryotic gDNA clones are absent from cDNA clones.

Library Construction

The nature of the cloned DNA is one aspect of library construction. The researcher also must choose an appropriate vector and host and decide how the library will be screened for the desired DNA fragment.

Nucleic acid

Once isolated, gDNA must be cleaved with restriction enzymes. The resulting fragments will range in size. Smaller fragments, though easier to clone into vectors, may not represent all possible genes. Some genes may be cut in the coding region, resulting in an incomplete gene sequence. This problem may be overcome by performing a partial digest of the gDNA to generate fragments of all sizes. Some may be too large to be cloned, so a second library, created by using a second restriction enzyme, might be necessary. When preparing a gDNA library, it is important to note that millions of individual cells have probably been used for DNA isolation. Therefore, many copies of any one gene should exist in the resulting library, some complete in sequence and some incomplete. Consider this: a gDNA library of the human genome (3 billion base pairs) would need to consist of 300,000 clones containing an average fragment length of 20,000 base pairs to include the whole sequence just one time. A gDNA library is essential if one wishes to study the function of regulatory DNA and intron sequences.

A cDNA library begins with isolation of mRNA. Since mRNA is produced only in tissues and cell types that transcribe and translate a particular gene, one must carefully consider the source of material used for its extraction. One would not use cells of the immune system as a starting point to obtain the gene for insulin, nor would one use pancreatic cells to obtain the gene for one of the interleukin molecules. When isolating RNA from cells, ribosomal RNA, transfer RNA, and mRNA are all present and extracted. Functional eukaryotic mRNA contains a poly(A) tail, a series of adenine residues at the 3′ end. Column affinity chromatography, using oligo(dT) (short strings of about 15 thymidine residues) bound to cellulose beads, captures mRNA by complementary base pairing. Transfer RNA and ribosomal RNA pass through the column. mRNA is eluted by breaking the A-T bonds. Once isolated, mRNA is transcribed into a DNA copy using the enzyme reverse transcriptase, producing an RNA-DNA hybrid. Next, the mRNA molecule is removed with ribonuclease (RNase) H. DNA polymerase is used to complete the second-strand synthesis. Then the resulting double-stranded DNA is ligated into the cloning vector.

Library cloning vectors

Cloning vectors for making DNA libraries include plasmids, lambda phage, cosmids, and yeast artificial chromosomes (YACs). One of the most important things to be considered when selecting a vector is the fragment size each will accept and replicate. Table 29.1 lists the average vector genome size and the maximum size of insert DNA accepted.

Table 29.1 Average vector size and maximum size of DNA inserts accepted

Vector	Genome size	Insert accepted	Origin
Plasmid	10 kb	10 kb	Bacterial chromosome
Lambda phage	50 kb	23 kb	Viral
Cosmid	4–6 kb	45 kb	Derived from lambda phage
YAC	11–12 kb	1,000 kb	Derived from yeast chromosomes

Plasmids occur naturally in bacteria and yeast and can range from 1 to 100 kilobases (kb) in size, though the average size is 10 kb. Recombinant DNA technology has produced many specialized plasmids engineered with multiple cloning sites and selectable markers. They can usually accept DNA inserts of only about 10 kb. For additional information about plasmids, see exercises 8, 9, 13, 14, 15, 20, and 21.

Lambda bacteriophage is probably the best-studied bacterial virus. Its genome is 50 kb and can accept inserts of up to 23 kb. Lambda contains cohesive overhangs (cos sites) that can pair and cause circularization of the genome in vivo. DNA can be cloned into lambda, which will then produce fully infectious recombinant virions. The infectious virus particles are about 1,000 times more efficient at transferring DNA into a host cell than is plasmid transformation. Many variations of lambda phage have been produced.

Cosmids are hybrids of two cos sites from lambda phage and about 5 kb of plasmid DNA. DNA inserts of up to about 45 kb can be accepted. Once the recombinant molecule is produced, it is mixed with a lambda packaging extract, a mixture of phage proteins, which packages naked DNA into phage heads in vitro, producing infective virions in test tubes. When cos sites are separated by 40 to 50 kb of insert DNA, the recombinant molecule will be packaged. Less than 38 kb and more than 52 kb will not produce an infective lambda virus particle. The recombinant lambda is used to infect *Escherichia coli* host cells. Once inside the host, the cosmid circularizes and replicates as a large plasmid.

YACs accept the largest DNA fragments, up to 1,000 kb (1 megabase). YACs are derived from yeast plasmids but replicate and segregate like linear chromosomes, because they include yeast origins of replication, centromeric sequences, and telomeric sequences. YACs are useful in constructing human gDNA libraries containing large fragments of chromosomes.

Library screening

Once a DNA library has been produced, the clone containing the desired DNA fragment must be isolated. This may be done by DNA hybridization on a nylon or nitrocellulose membrane (similar to Southern blotting) or by examining clones for protein expression. Both methods require some information about the gene sought. Probes may be derived from known DNA of closely related organisms, or synthesized in the laboratory based on information about the amino acid sequence of the desired gene.

To screen a library using DNA hybridization, colonies of yeast or bacteria or phage plaques on a bacterial lawn grown on solid medium in a petri dish are transferred to a nylon or nitrocellulose membrane by laying it on top of the colonies. When the membrane is lifted off, it is coated with a replica of the microbial colonies and the recombinant DNA molecules they contain. Some of the microbial colony remains on the agar, however. Colonies on the membrane are lysed, and the DNA molecules are denatured and hybridized to DNA, RNA, or short synthetic oligonucleotide probes. The excess probe is washed off, and bound complementary probe is detected. The hybridization process is essentially the same as in a Southern blot (exercise 16). Colonies containing recombinant clones complementary to the probe appear as a dot on the filter. Since the filter is a replica of the original plate, the location of the colony containing the desired insert DNA can be identified and the clone subcultured. The colony is grown in larger volumes of medium, tested again to be sure the DNA of interest is present, then used to replicate large quantities of the desired DNA.

A second method for screening a library is to assay for the expressed protein. This can be done phenotypically by looking for visual expression of a trait or for enzyme activity. Expressed protein may also be detected using an immunological assay. Antibody screening is carried out in a manner similar to Western blotting (exercise 19), using nitrocellulose filter lifts of colonies as described above. Colonies of microbes on the filter are lysed and then probed with a primary antibody that binds to the protein of interest. The antibody is detected with a secondary antibody, one that binds to the primary antibody and is visualized in a colorimetric enzyme assay. Positive spots, correlated to the master plate, are used to pick colonies containing recombinant clones of interest.

In this laboratory exercise, we will make a genomic library of *Erwinia herbicola* DNA. *E. herbicola* is a yellow-pigmented gram-negative bacillus in the family *Enterobacteriaceae*. This species is a saprophyte, often colonizing plant surfaces. *E. herbicola* seems to produce antimicrobial substances that attack other microorganisms on the plant surface. This ability has led ecologists to use *E. herbicola* as a biological control agent to prevent plant diseases. It is believed that *E. herbicola* can survive on the leaf surface owing to carotenoid production, which provides resistance to ultraviolet radiation. Recent studies show that carotenoid synthesis by *E. herbicola* Eho 10, our experimental organism, is encoded by carotenoid genes of genomic origin.

Our cloning vector is the cosmid pWE15 (Fig. 29.1), an 8.2-kb vector, which contains bacteriophage promoter sequences flanking a unique cloning site for *Bam*HI. pWE15 carries the lambda cos site for lambda packaging, the Amp^r gene, and a bacterial origin of replication for selection and replication in *E. coli*. In addition, it carries the SV40 origin and the neomycin marker for expression and selection in eukaryotic cells.

We will transform *E. coli* with our recombinant cosmid pWE15, containing the *E. herbicola* restriction fragments, thereby creating the gDNA library. The library will be screened by observing phenotypic expression. If genes for carotenoid synthesis are present in a particular clone, the colony will be yellow.

Lab Language

cDNA
cos site
cosmid
DNA library
exon
expression library
gDNA
intron
lambda phage
packaging extract
plasmid
reverse transcriptase
YAC

Laboratory Safety

Review the sections on physical, biological, and chemical hazards in "Laboratory Safety." Special considerations for this exercise:

- use of microbes
- Bunsen burner
- warm or hot solutions
- guanidium HCl
- chloroform

Laboratory Sequence

Day 1: Isolation and quantitation of *E. herbicola* gDNA (protocol 1)
Day 2: Partial restriction enzyme digest of gDNA (protocol 2)

Figure 29.1 Cosmid pWE15.

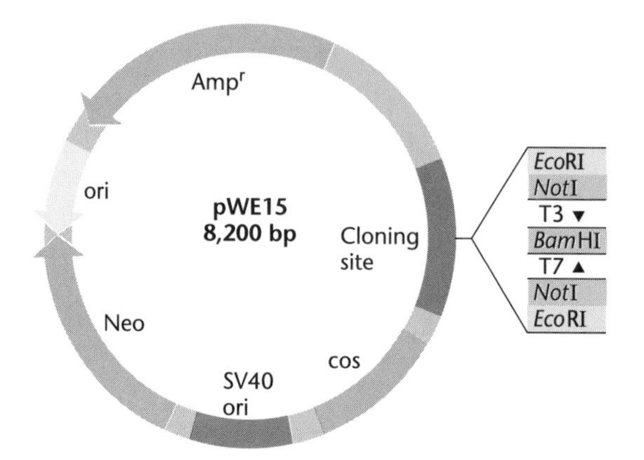

Linearization and purification of pWE15 (protocol 3)

Day 3: Dephosphorylation and purification of pWE15 (protocol 3)
Ligation reaction (protocol 4) (overnight)
Start XL-1 Blue MR culture (protocol 5) (overnight)

Day 4: Packaging of library DNA (protocol 6)
Infection and amplification of the cosmid library (protocols 5 and 7) (overnight)

Day 5: Titer of the amplified library (protocols 7 and 8) (overnight)

Day 6: Observation and analysis of results

Materials

Standard Materials for the Exercise

0.5- to 10-µl micropipette and sterile tips
10- to 100-µl micropipette and sterile tips
100- to 1,000-µl micropipette and sterile tips
microcentrifuge
sterile microcentrifuge tubes
wax or permanent marking pen
gloves
microcentrifuge tube rack
beaker for collection of biological waste

Isolation and Preparation of *E. herbicola* gDNA

Class equipment and supplies

overnight *E. herbicola* broth culture
Wizard gDNA Purification System (Promega)
- nuclei lysis solution (Tris-EDTA-sodium dodecyl sulfate)
- RNase A solution
- protein precipitation solution (ammonium acetate)
- rehydration solution or 1× Tris-EDTA (TE)

100% isopropanol
70% ethanol
37°C water bath or heat block
80°C water bath or heat block
65°C water bath or heat block
*Sau*3AI restriction enzyme and 10× buffer
ice

Materials per pair of students

1 aliquot of nuclei lysis solution
1 aliquot of RNase A
1 aliquot of protein precipitation solution
1 aliquot of 100% isopropanol
1 aliquot of 70% ethanol
1 aliquot of hydration solution
1 aliquot of *Sau*3AI and 10× buffer
Pasteur pipette and bulb

Preparation of the Cosmid Vector pWE15

Class equipment and supplies

37°C water bath or heat block
65°C water bath
*Bam*HI and 10× buffer
Wizard DNA Clean-up System (Promega)
- cleanup resin in guanidine HCl
- minicolumn
- vacuum manifold

calf intestinal alkaline phosphatase (CIAP) and buffer
0.5 M EDTA
80% isopropanol

Materials per pair of students

1 aliquot of *Bam*HI and 10× buffer
1 aliquot of cleanup resin
1 aliquot of 0.5 M EDTA

Ligation of the Library DNA

Class equipment and supplies

T4 DNA ligase and buffer
refrigerator

Students will need to prepare

linearized and dephosphorylated plasmid (protocol 4)
partially digested gDNA (protocol 2)
overnight culture of XL-1 Blue MR *E. coli*

Materials per pair of students

1 aliquot of T4 DNA ligase
1 aliquot of 10× DNA ligation buffer
sterile dH_2O

EXERCISE 29

Packaging, Amplification, Titering, and Identification of Carotenoid Clones

Class equipment and supplies
37°C bacterial shaker
Luria-Bertani (LB) broth
37°C incubator
Gigapack III XL packaging extract
suspension medium (SM) buffer
chloroform
30°C bacterial shaker
10 mM $MgSO_4$
spectrophotometer and quartz cuvettes
10-mg/ml ampicillin

Students will need to prepare
LB agar + ampicillin

Materials per pair of students
10 LB agar + ampicillin petri dishes
inoculating loop
Bunsen burner
spreading rods
20-ml liquid LB medium

Protocols

PROTOCOL 1 gDNA EXTRACTION FROM GRAM-NEGATIVE BACTERIA

This strain of bacteria has been characterized by the American Type Culture Collection as an atypical (sorbitol-positive) strain of Escherichia vulneris, *and DNA hybridization concurs; the original nomenclature will be used, however.*

Exercise 22 uses a slightly different method for extraction of gDNA from bacteria; both work well.

1. Obtain an overnight broth culture of *E. herbicola.*
2. Label a microcentrifuge tube. Transfer approximately 1.0 ml of the culture into the tube, using a Pasteur pipette.
3. Centrifuge the tube at 14,000 rpm for 2 minutes to pellet the bacteria.
4. Carefully remove most of the supernatant by pouring it into a container of disinfectant.
5. Add 800 µl of nuclei lysis solution and gently pipette up and down to resuspend cells.
6. Incubate at 80°C for 5 minutes.

The sample must be cool; heat will denature the RNase enzyme.

7. Cool the sample at room temperature and add 5 µl of RNase A solution to the cell lysate. Mix by inverting the tube 25 times.
8. Incubate for 15 to 60 minutes at 37°C.

Ammonium acetate causes the proteins to precipitate, leaving the DNA in solution.

9. Cool the sample at room temperature. Add 300 µl of protein precipitation solution and vortex vigorously for 20 seconds.
10. Incubate on ice for 3 minutes.
11. Microcentrifuge the sample at 14,000 rpm for 3 minutes to pellet the protein debris.
12. Pipette or pour the supernatant into a clean 1.5-ml tube, leaving the protein debris behind.

Vigorous mixing will shear DNA into small fragments.

13. Add 800 µl of 100% isopropanol to the supernatant and mix *gently* by inverting the tube at least 50 times.

The pellet may appear off-white to yellow.

14. Centrifuge at 14,000 rpm for 2 minutes to pellet the DNA.
15. Pour off the supernatant and drain the liquid on an absorbent towel.

16. Add 800 μl of 70% ethanol and invert the tube several times to wash the pelleted DNA.

A 70% ethanol wash removes unwanted salts from the DNA.

17. Centrifuge at 14,000 rpm for 2 minutes to pellet the DNA.

18. Pour off the supernatant, *slowly and carefully,* watching the pellet so it does not pour out.

19. Invert the tube and drain all of the ethanol. Air dry the DNA pellet for at least 30 minutes.

Any residual ethanol will interfere with subsequent analysis.

20. Add 80 μl of rehydration solution or 1× TE to the pellet.

21. Incubate in a 65°C water bath or heat block for 1 hour. Mix occasionally by tapping the tube.

22. Quantitate the concentration of isolated DNA by spectrophotometry (exercise 11) or gel electrophoresis (exercise 12).

23. Adjust the concentration, if necessary, to between 0.6 and 1.0 μg/μl.

PROTOCOL 2 PARTIAL RESTRICTION ENZYME DIGESTION OF gDNA

1. Set up the following digestion using the gDNA isolated in protocol 1.

   43 μl of gDNA (0.6 to 1.0 μg/μl)
   2 μl of *Sau*3AI restriction enzyme (10 U/μl)
   5 μl of 10× restriction enzyme buffer
   50-μl total

Sau*3AI recognizes the sequence:*
GATC
CTAG

Bam*HI-digested DNA has compatible ends.*

2. Incubate the digestion at 37°C for 40 minutes.

3. Heat-inactivate the enzyme by incubation at 65°C for 20 minutes.

4. Use the *Sau*3AI-digested gDNA for library construction.

PROTOCOL 3 PREPARATION OF LIBRARY CLONING VECTOR

A. Linearization of pWE15 with *Bam*HI

1. Digest 4.0 μg of pWE15 with 70 units of *Bam*HI in a total volume of 30 μl. Document your digestion below.

Excess enzyme and long incubation times ensure complete digestion.

Bam*HI recognizes the sequence:*
↓
GGATCC
CCTAGG
↑

2. Incubate the digestion at 37°C for 2 hours.

B. Purification of the Linearized pWE15

1. Warm and maintain 1× TE at 65°C.

Exercises 13 and 14 use a matrix system to isolate DNA fragments; all work well.

2. Obtain a 3-ml syringe barrel and minifilter. Attach the filter to the barrel and insert the assembly into the vacuum manifold.

3. Add 20 μl of dH_2O to the restriction enzyme digestion from protocol 3A. Pop spin to mix and concentrate.

4. Mix the Wizard cleanup resin. Pipette 1 ml of the resin solution into the digested pWE15 sample and mix by inverting the tube.

The resin, with bound DNA, will not pass through the filter.

5. Pipette the resin and DNA mixture into the syringe barrel. Apply vacuum to draw the liquid through the filter.

6. Add 2 ml of 80% isopropanol to the syringe barrel and apply vacuum to dry the liquid through the filter.

Do not overdry the resin.

7. Apply vacuum for an additional 30 seconds after the liquid has been pulled through to dry the resin.

8. Remove the minifilter from the vacuum manifold and syringe barrel. Place the filter into a clean microcentrifuge tube.

Isopropanol contamination will interfere with subsequent steps and DNA manipulations.

9. Microcentrifuge the filter at 14,000 rpm for 2 minutes to remove residual isopropanol.

10. Remove the minifilter to a clean microcentrifuge tube.

11. Add 30 µl of prewarmed (65°C) 1× TE.

12. Incubate for 3 minutes at room temperature.

13. Centrifuge the minifilter at 14,000 rpm for 20 seconds to elute the bound DNA from the resin.

C. Dephosphorylation of pWE15

CIAP removes 5′-phosphate groups, preventing the vector from recircularizing.

1. Set up the dephosphorylation reaction:

   30 µl of linear, purified WE15
   4 µl of CIAP (1 U/µl)
   4 µl of 10× CIAP reaction buffer
   2 µl of dH_2O
   40-µl total

2. Pop spin to mix and concentrate reagents.

3. Incubate at 37°C for 60 minutes.

4. Add 2 µl of 0.5 M EDTA to stop the reaction.

D. Purification of the Dephosphorylated pWE15 Vector

Enzyme molecules must be removed from the ends of the DNA for ligation to occur.

Repeat the purification protocol in protocol 3B to remove CIAP enzyme from the pWE15 vector. *Do not add dH_2O as instructed in step 3.*

PROTOCOL 4 LIGATION OF *E. HERBICOLA* gDNA AND pWE15 COSMID VECTOR

It is estimated that 1 to 1.5 µg of gDNA is in the ligation reaction.

1. Set up the following ligation reaction.

   15.7 µl of *E. herbicola Sau*3AI-digested gDNA
   10.0 µl of pWE15 vector (~0.1 µg/µl)
   4.0 µl of T4 DNA ligase (2 U/µl)
   3.3 µl of 10× ligation buffer
   33.0-µl total

2. Pop spin to mix and concentrate the reagents.
3. Incubate the ligation reaction overnight at 4°C.

PROTOCOL 5 PREPARATION OF XL-1 BLUE MR *E. COLI* FOR INFECTION BY PACKAGED LIBRARY DNA

1. Inoculate 50 ml of LB medium with one colony of *E. coli* XL-1 Blue MR.
2. Incubate the culture overnight at 30°C and shake.

30°C incubation prevents overgrowth.

3. Centrifuge the cells at 500 × *g* for 10 minutes to pellet the cells.
4. Remove the supernatant to disinfectant solution.
5. Add 15 ml of sterile 10 mM $MgSO_4$.

$MgSO_4$ promotes viral receptor expression.

6. Determine the optical density at 600 nm (OD_{600}) of the resulting culture.
7. Add sterile 10 mM $MgSO_4$ until the $OD_{600} = 1$.

PROTOCOL 6 PACKAGING OF THE LIBRARY DNA

1. Remove one tube of packaging extract from the freezer and place on ice until ready for use.

Packaging extract contains phage proteins necessary for DNA packaging in phage heads.

2. Begin thawing the tube by holding it between your fingers.
3. When the contents begin to thaw, immediately add 4 µl of the ligated DNA from protocol 4.
4. Pop spin to mix and concentrate the reagents.
5. Incubate at room temperature for 2 hours.
6. Add 500 µl of SM buffer.
7. Add 50 µl of chloroform to the tube and pop spin to mix.
8. Centrifuge the tube at 14,000 rpm for 2 minutes to pellet the debris.
9. Use the packaged library to infect the XL-1 Blue *E. coli* in protocol 7.

PROTOCOL 7 AMPLIFICATION OF THE COSMID LIBRARY

1. In a 15-ml sterile test tube, mix the packaged DNA from protocol 6 with an equal volume of $MgSO_4$-treated XL-1 Blue MR *E. coli* from protocol 5.
2. Incubate at room temperature for 30 minutes.

Phage attach to and infect E. coli.

3. Add 4 volumes of liquid LB medium.
4. Incubate at 37°C for 60 minutes and shake.

DNA replication begins.

5. Add 500 µl of liquid LB medium and mix.
6. Obtain five LB agar + ampicillin petri dishes. Pipette 100 µl of the cell suspension onto each plate.
7. Incubate the plates, upside down, overnight at 37°C.

8. After overnight incubation, add 3 ml of liquid LB medium to each plate.

9. Using a sterile inoculating loop, gently scrape all the cells off the plates and combine the suspensions in a sterile 50-ml test tube.

10. Repeat step 9, adding another 3 ml of liquid LB medium to each plate. Combine all liquid cell suspensions from all five plates in one 50-ml test tube.

11. Add 3 ml of liquid LB medium containing 100-μg/μl ampicillin.

12. Incubate at 37°C for 30 minutes and shake.

13. Use immediately or mix with glycerol storage buffer and freeze for future use.

PROTOCOL 8 TITER OF THE AMPLIFIED LIBRARY

1. Prepare 10 ml of liquid LB medium containing 100-μg/μl ampicillin using a 10-mg/ml stock of ampicillin. Document your calculations.

2. Select an aliquot of library-containing cells.

3. Prepare a serial dilution series of the library-containing cells by mixing 100 μl of the cells with 900 μl of LB + ampicillin. Make the following dilutions, using the previous tube as a source of the cells for each sequential 1:10 dilution. *Do not forget to mix each new dilution before removing cells.*

   1:10
   1:100
   1:1,000
   1:10,000
   1:100,000
   1:1,000,000

4. Obtain five LB agar + ampicillin plates and label each for a serial dilution. You will plate 10^{-2} through 10^{-6} dilutions.

5. Pipette 100 μl of each serial dilution onto the appropriately labeled LB agar + ampicillin plate.

6. Spread the cells evenly across the agar surface by gently moving the spreader back and forth while turning the petri dish.

7. Incubate the petri dishes upside down at 37°C overnight.

8. Count the colonies growing on each plate and determine the colony-forming units per milliliter (CFU/ml).

9. Observe the colonies for those with a yellow color. These are the clones containing the genes for carotenoid production.

References and General Reading

Ausubel, F. M., R. Brent, R. E. Kingston, D. D. Moore, J. G. Seidman, J. A. Smith, and K. Struhl (ed.). 1994. *Current Protocols in Molecular Biology.* Wiley Interscience, New York, N.Y.

Beer, S. V., J. R. Rundle, and J. L. Norelli. 1984. Recent progress in the development of biological control for the fire blight—a review. *Acta Hortic. (The Hague)* **151:**195–201.

Doyle, K. (ed.). 1996. *Protocols and Applications Guide,* 3rd ed. Promega, Madison, Wis.

Glick, B. R., and J. J. Pasternak. *Molecular Biotechnology: Principles and Applications of Recombinant DNA,* 2nd ed. ASM Press, Washington, D.C., 1998.

Misawa, N., M. Nakagawa, K. Kobayashi, S. Yamano, Y. Izawa, K. Nakamura, and K. Harashima. 1990. Elucidation of the *Erwinia uredovora* carotenoid biosynthetic pathway by functional analysis of gene products expressed in *Escherichia coli. J. Bacteriol.* **172:**6704–6712.

Perombelon, M. C. M. 1992. The genus *Erwinia,* p. 2899–2921. *In* A. Balows, H. G. Truper, M. Dworkin, W. Harder, and K. H. Schleifer (ed.). *The Prokaryotes,* 2nd ed. Springer-Verlag, New York, N.Y.

Perry, K. L., T. A. Simonitch, K. J. Harrison-Lavoie, and S. T. Liu. 1986. Cloning and regulation of *Erwinia herbicola* pigment genes. *J. Bacteriol.* **168:**607–612.

Rosenthal, N. 1994. Molecular medicine: stalking the gene-DNA libraries. *N. Engl. J. Med.* **331:**599–600.

Stratagene 1998. *Technical Manual for Gigapack II Gold Packaging Extract, GigapackIII Plus Packaging Extract, and Gigapack III XL Packaging Extract.* Stratagene, La Jolla, Calif.

Stratagene. 1998. *Technical Manual for pWE15 Vector.* Stratagene, La Jolla, Calif.

Turner, P. C., A. G. McLennan, A. D. Bates, and M R. H. White. 1997. *Instant Notes in Molecular Biology.* BIOS Scientific Publishers, Springer-Verlag, New York, N.Y.

Winkelman, G., R. Lupp, and G. Jung. 1980. Herbicolins. New peptide antibiotic from *Erwinia herbicola. J. Antibiot.* (Tokyo) **33:**353–358.

Discussion

1. Determine the number of clones required to contain the *E. coli* genome (4 million base pairs) one time if each fragment inserted averages 2,000 base pairs in length.
2. Compare and contrast the use of gDNA and cDNA in library construction.
3. Why is a partial restriction enzyme digestion performed on the DNA used for library construction? Discuss how this is done and any limitations of using this DNA.
4. List and discuss advantages and disadvantages of the various vectors that may be used for cloning libraries.
5. Describe and discuss the purpose(s) of packaging the DNA library in protocol 6.
6. DNA libraries may be screened with nucleotide probes or by looking for protein expression. Discuss the advantages and disadvantages of each and when one method may be preferred.
7. Determine the CFU/ml of your stock DNA library. Show your calculations.
8. What is the frequency of clones in your library that express the genes for carotenoid synthesis?
9. Are these the only clones in your library that contain genes for carotenoid synthesis? Explain your answer.

Appendix

The field of biotechnology has grown so much, and the applications have been so widely used over the past few years, that virtually any solution and reagent can probably be purchased ready to use. This can greatly speed up experiments, ensure quality control, and reduce the risks associated with manipulation of certain hazardous and toxic chemicals. One must pick and choose carefully what is purchased and the grade of reagent obtained. Some of the formulations listed below may vary slightly from one laboratory to another.

Below are listed common formulations, methodologies for preparing reagents, oligonucleotide sequences, a list of suppliers, and other information. No one supplier is recommended over another; you may find one that suits your needs better than these. Listed are specific supplies and reagents that we have used. Many more are available and appropriate. One should comparison shop!

Common Formulations

2% Acridine Orange

1. Dissolve in dH_2O.
2. Store protected from light.

10% Ammonium Persulfate (APS)

1. Dissolve wt/vol in dH_2O.
2. Prepare fresh each week or store aliquots at −20°C.

Ampicillin

1. Dissolve 10 mg/ml in dH_2O.
2. Filter sterilize (or use sterile dH_2O).
3. Store aliquots at −20°C.

BCIP-NBT (5-Bromo-4-Chloro-3-Indolylphosphate–Nitroblue Tetrazolium)

BCIP stock

1. 100 mg of BCIP in 2 ml of 100% dimethylformamide (DMF)
2. Store at 4°C for ~1 year or at −20°C.

NBT stock

1. 100 mg of NBT in 2 ml of 70% DMF.
2. Store at 4°C for ~1 year or at −20°C.

Note: Premixed solutions can be purchased from Bio-Rad.

NBT/BCIP developing solution

33 µl of NBT stock
5 ml of NBT/BCIP developing buffer
17 µl of BCIP stock
Stable for 1 hour at room temperature.

BCIP/NBT developer buffer
(alkaline phosphate substrate buffer)

100 mM Tris HCl, pH 9.5
100 mM NaCl
5 mM $MgCl_2$

Note: Premixed buffer comes with NBT/BCIP from Bio-Rad.

Blotting Buffer—for Western blotting

1× Tris-glycine
20% methanol

1. Adjust volume with dH_2O.
2. Store at room temperature.

2-mg/ml Bovine Serum Albumin (BSA) Fraction V

Dissolve wt/vol in 0.9% NaCl.

10-mg/ml BSA

Dissolve wt/vol in 1× physiologically buffered saline (PBS).

1.0 M $CaCl_2$ (Calcium Chloride)

1. Dissolve in dH_2O.
2. Autoclave to sterilize.

Calf Thymus DNA

10 mg/ml in 1× Tris-EDTA (TE) or 1× Tris-sodium chloride-EDTA (TNE)

2-mg/ml Single-Stranded Carrier DNA (salmon sperm DNA, Sigma #D1626)

1. Dissolve 200 mg of DNA in 100 ml of sterile 1× TE.
2. Disperse by pipetting up and down.
3. Mix vigorously using a stir plate for 2 to 3 hours or overnight in the cold.
4. Aliquot and store at −20°C.
5. Boil for 5 minutes and chill quickly on ice before use.

Chromatography Solvent—for detection of nopaline

120 ml of formic acid
700 ml of isopropanol
180 ml of dH_2O

Coomassie Blue Stain

25% isopropanol
10% acetic acid
0.5% wt/vol Coomassie blue R-250

1. Adjust volume with dH_2O.
2. Store at room temperature.
3. May be reused.

DNA Loading Dye

50% glycerol
0.25% bromophenol blue
0.25% xylene cyanol

Destaining Solution—for sodium dodecyl sulfate (SDS)-polyacrylamide gel electrophoresis

7.5% methanol
7.5% acetic acid

Denaturation Buffer—for Southern blotting

1.5 M NaCl
0.5 M NaOH

0.5 M EDTA, pH 8.0

1. Dissolve 93.05 g in 300 ml of dH_2O.
2. Adjust pH to 8.0 with concentrated NaOH or NaOH pellets.
3. The EDTA will go into solution as pH reaches 8.0.
4. Adjust volume to 500 ml with dH_2O.
5. Autoclave to sterilize.
6. Store at room temperature.

10-mg/ml Ethidium Bromide Solution

1. Dissolve in dH_2O.
2. Store at 4°C protected from light.

Note: We suggest purchasing premixed solutions to minimize exposure.

Equilibration Buffer—for Southern blotting

0.10 M Tris, pH 9.5
0.10 M NaCl
0.05 M $MgCl_2$

2× GSB (Glycerol Storage Buffer)—for storing bacterial stocks

25 mM Tris, pH 8.0
0.1 M $MgSO_4$
65% glycerol

Prepare with sterile reagents and sterile dH_2O.

GTE (Glucose-Tris-EDTA) Solution—for alkaline miniprep

50 mM glucose
25 mM Tris, pH 8.0
10 mM EDTA

Homogenates

1% chicken liver homogenate
1% egg yolk homogenate
1% egg white homogenate
1% fish egg (caviar) homogenate
1% ovalbumin
1% tofu (firm) homogenate

1. Blend 1 g in 100 ml of 0.9% NaCl.

1:10 (10%) egg yolk
1:10 (10%) egg white

1. Separate yolk from white.
2. Mix 1 ml of each with 9 ml of 1× PBS.

IKI (Iodine–Potassium Iodide) Solution

1. Dissolve 3 g of potassium iodine in 25 ml of dH_2O.
2. Add 0.6 g of iodine and stir until dissolved.
3. Adjust volume to 200 ml with dH_2O.
4. Store in a dark bottle (solution is light sensitive) at room temperature.

1 M IPTG (Isopropyl-β-D-Thiogalactopyranoside)

1. Dissolve 1.2 g in 5 ml dH_2O
2. Filter sterilize or use sterile dH_2O.
3. Store at 4°C or store aliquots at −20°C.

2× Laemmli Sample Buffer (LSB)

80 mM Tris, pH 6.8
2% SDS
100 mM dithiothreitol
20% glycerol
0.01% bromophenol blue

Note: Premixed solutions are available from Bio-Rad.

1 M Lithium Acetate (LiAc)

1. Prepare in dH_2O.
2. Adjust pH to 8.5.
3. Autoclave.
4. Store at room temperature.

Ninhydrin Reagent

0.2 g of ninhydrin dissolved in 100 ml of 95% ethanol

0.5 M $MgCl_2$ (Magnesium Chloride)

1. Prepare in dH_2O.
2. Autoclave.
3. Store at room temperature.

1 M $MgSO_4$ (Magnesium Sulfate)

1. Prepare in dH_2O.
2. Autoclave.
3. Store at room temperature.

3 M NaOAc (Sodium Acetate), pH 4.8

1. Dissolve 40.8 g of sodium acetate · $3H_2O$ in H_2O.
2. Adjust pH with 3 M acetic acid.

Neutralization Buffer—for Southern blotting

1.5 M NaCl
0.5 M Tris, pH 7.0

NS (Sodium Hydroxide–SDS) Solution—for alkaline miniprep

0.2M NaOH
1% SDS

0.5-µg/µl Papain (Worthington Biochemical Corp., #LS003124)

1. Dissolve in dH_2O.
2. Store aliquots at −20°C.

10-mg/ml Ovalbumin

1. Dissolve wt/vol in 1× PBS.
2. Store aliquots at −20°C.

10× PBS

80.0 g of NaCl
2.0 g of KCl
11.5 g of $Na_2HPO_4 \cdot 7H_2O$
2.0 g of KH_2PO_4

1. Add the powdered reagents to dH_2O in a graduated cylinder, with stirring.
2. Adjust the volume to 1 liter.
3. May be autoclaved.
4. Store at room temperature.

1× PBS-T

1× PBS + 0.5% Tween 20

PCR-H Buffer—for polymerase chain reaction (PCR) of DNA extracted from a hair follicle

50 mM KCl
10 mM Tris-HCl, pH 8.3
2.5 mM $MgCl_2$
0.1-mg/ml gelatin
0.45% NP-40
0.45% Tween 20

Polyethylene Glycol (PEG 50% wt/vol)

1. Add 50 g of PEG to 35 ml of dH_2O in a 100-ml glass beaker.
2. Stir until dissolved, about 30 minutes.
3. Transfer the liquid to a 100-ml graduated cylinder.
4. Rinse the beaker with a small amount of dH_2O and add this to the graduated cylinder.
5. Adjust the volume to 100 ml.
6. Mix well by inversion.
7. Autoclave.
8. Store in a securely capped bottle at room temperature.

Note: The PEG concentration, 50%, is very important to ensure transformation efficiency. Evaporation from the PEG stock solution will result in an increase in the concentration of PEG and will severely reduce the efficiency.

Prehybridization-Hybridization Buffer—for Southern blotting

5× SSC
5% blocking reagent (BSA)
0.1% sarcosine
0.02% SDS
50% formamide

Note: Formamide can be deleted if hybridization temperature is raised.

10-mg/ml Proteinase K

1. Dissolve in sterile dH_2O.
2. Aliquot and store at −20°C.

20-mg/ml RNase A (DNase free)

1. Dissolve 1 mg/ml in 1× TE.
2. Boil 10 to 30 minutes.
3. Cool slowly to room temperature.
4. Store aliquots at −20°C.

Note: DNase-free solutions may be purchased.

10% SDS

1. Dissolve 10 g of SDS in 80 ml of warm (65°C) dH_2O.
2. Adjust volume to 100 ml.

Do not shake!

10× SDS Running Buffer—for SDS-polyacrylamide gel electrophoresis

30.0 g of Tris base
144.4 g of glycine
10.0 g of SDS

1. Mix Tris and glycine in dH_2O.
2. Adjust volume to 1 liter.
3. Add SDS and mix gently.
4. Store at room temperature.

20× SSC

3.0 M NaCl
0.3 M sodium citrate

Adjust pH to 7.0 with 10 M NaOH.

10× TAE (Tris-Acetate-EDTA)

48.4 g of Tris base
11.4 ml of glacial acetic acid
7.44 g of $Na_2EDTA \cdot 2H_2O$

Adjust volume to 1,000 ml with dH_2O.

10× TBE (Tris-Borate-EDTA)

100 g of Tris base
55 g of boric acid
40 ml of 0.5 M EDTA, pH 8.0

1× TE

10 mM Tris, pH 7.4
1 mM EDTA, pH 8.0

1 M Tris Buffers

1. Dissolve 121 g of Tris base in 800 ml of dH_2O.
2. Adjust to desired pH with concentrated HCl.
3. Adjust volume to 1 liter.
4. Store at room temperature.

Note: The pH of Tris buffers changes with temperature, decreasing about 0.028 pH units per degree Celsius.

1.5 M Tris, pH 8.8

1. Dissolve 181.5 g of Tris base in 700 ml of dH_2O.
2. Adjust to desired pH with concentrated HCl.
3. Adjust volume to 1 liter.
4. Store at room temperature.

Note: The pH of Tris buffers changes with temperature, decreasing about 0.028 pH units per degree Celsius.

10× Tris-Glycine—for blotting buffer

30.3 g of Tris base
143.0 g of glycine

1. Mix Tris and glycine in dH_2O.
2. Adjust volume to 1 liter.

10× TNE

100 mM Tris base
2.0 M NaCl
10 mM EDTA

1. Dissolve in dH_2O.
2. Adjust pH to 7.4 with concentrated HCl.
3. Store at room temperature.

0.9% (Isotonic) NaCl

1. Dissolve in dH_2O.
2. Autoclave.
3. Store at room temperature.

X-Gal (5-Bromo-4-Chloro-3-Indolyl-β-D-Galactoside)

1. 20 mg/ml dissolved in DMF
2. Store aliquots at −20°C.

Microbial Media

(May be purchased as premixed powders or premade solutions.)

LB (Luria-Bertani) Broth

10 g of tryptone
5 g of yeast extract
5 g of NaCl
1 ml of 1 M NaOH

1. Makes 1 liter.
2. Autoclave to sterilize.

LB Agar

Add 15 g of agar per liter of LB medium before autoclaving.

3-ml Vials of LB Agar

Dispense 3 ml of hot, sterile agar into sterile vials.

LB + Ampicillin + Starch Plates

1. Prepare 200 ml of LB medium.
2. Add 2% (wt/vol) starch.
3. Autoclave.
4. Cool and add ampicillin.
5. Pour a thin layer of LB + ampicillin + starch over LB + ampicillin plates.
6. Allow to harden, then store wrapped at 4°C.

Murashige and Skoog (MS) Media

Macro salts

1.65 g of NH_4NO_3
1.90 g of KNO_3
0.44 g of $CaCl_2 \cdot 2H_2O$
0.37 g of $MgSO_4 \cdot 7H_2O$
0.17 g of KH_2PO_4

Micro salts

27.80 mg of $FeSO_4 \cdot 7H_2O$
33.60 mg of $Na_2EDTA \cdot 2H_2O$
0.83 mg of KI
6.20 mg of H_3BO_4
22.30 mg of $MnSO_4 \cdot 4H_2O$
8.60 mg of $ZnSO_4 \cdot 7H_2O$
0.25 mg of $Na_2MoO_4 \cdot 2H_2O$
0.025 mg of $CuSO_4 \cdot 5H_2O$
0.025 mg of $CoCl_2 \cdot 6H_2O$

Organic supplements

100.00 mg of myoinositol
0.05 mg of nicotinic acid
0.05 mg of pyrodoxine HCl
0.05 mg of thiamine HCl
0.20 mg of glycine
20.00 g of sucrose

1. Dissolve the components in ~800 ml of dH_2O.
2. Adjust pH to 5.7 with 1 M NaOH.
3. Adjust volume to 1,000 ml with dH_2O.
4. Autoclave to sterilize.

Synthetic Complete (SC) Dropout Media

Prepare solutions A, B, and C.

Solution A—Yeast Base

1. Dissolve 6.7 g of yeast nitrogen base, without amino acids (Difco #0919-15), in 800 ml of dH_2O.
2. Adjust pH to 5.8 with 10 M NaOH.
3. Autoclave and cool to 65°C.

For plates, add 20 g of agar before autoclaving.

Note: Volume will be adjusted with the addition of solutions B and C.

Solution B—20% Galactose Solution

1. Dissolve 20 g of galactose in 80 ml of dH_2O.
2. Stir to dissolve galactose.
3. Adjust volume to 100 ml.
4. Sterilize by filtration.

Solution C—10× Dropout Mixture

333 mg of adenine sulfate
333 mg of uracil
333 mg of L-tryptophan
333 mg of L-histidine-HCl
333 mg of L-arginine-HCl
333 mg of L-methionine
500 mg of L-tyrosine
1,000 mg of L-leucine
500 mg of L-lysine-HCl
833 mg of L-phenylalanine
1,666 mg of L-aspartic acid
3,333 mg of L-threonine

1. Dissolve components in 400 ml of dH_2O.
2. Stir until all components are dissolved.
3. Adjust the volume to 500 ml.
4. Sterilize by filtration.

SC Medium Preparation

1. Warm solution B and solution C to 50°C.
2. Mix 800 ml of solution A with 100 ml of solution B and 100 ml of solution C.
3. Flick your medium bottle to mix the medium components.
4. If making plates, pour plates.

SC Dropout Media for Selection of Transformants

Solution D—10× dropout mixture without uracil

Solution E—10× dropout mixture without tryptophan

Solution F—10× dropout mixture without uracil and tryptophan

1. Prepare SC dropout media as solution C, leaving out the respective component(s).
2. Prepare the three types of media, using solution D, E, or F instead of solution C.

Suspension Medium (SM Buffer)

5.8 g of NaCl
2.0 g of $MgSO_4 \cdot H_2O$
50 ml of 1 M Tris-HCl (pH 7.5)
5 ml of 2% gelatin

1. Mix in dH_2O.
2. Adjust volume to 1,000 ml.

Antibodies

1° Antibodies

antiovalbumin (Accurate Chemical or Rockland Immunochemicals)

antipapain (Rockland Immunochemicals #100-1174)

anti-human α-amylase (salivary or pancreatic) (Accurate Chemical or Rockland Immunochemicals)

anti-bacterial amylase (Rockland Immunochemicals)

anti-β-amylase (Rockland Immunochemicals)

2° Antibodies

Make sure the 2° antibody is specific for the species for 1° antibody and alkaline phosphatase labeled.

Accurate Chemicals, Bio-Rad, Boehringer Mannheim, Promega, Rockland Immunochemicals, and others

Restriction Enzymes and Other Biotechnology Enzymes

New England BioLabs, Promega, MBI Fermentas, and others

Microbe Strains

Bacteria

Escherichia coli DH5α

E. coli JM109 (Promega)

E. coli JM109 (DE3) (Promega)

Agrobacterium tumefaciens (ATCC #33970)

E. coli XL-1 Blue MR (Stratagene) (with pWE15)

Erwinia herbicola (listed as *Escherichia vulneris*) (ATCC #39368)

The *E. coli* strains used in biotechnology are very common, and many companies supply and sell them.

Yeast

Saccharomyces cerevisiae YPH499 (ATCC #76625)

Other

Euglena gracilis (Carolina Biologicals)

Spirogyra sp. (Carolina Biologicals)

DNA

pGFPuv (Clontech #6079-1)

pGEM-3Zf+ (Promega #P2271)

pHGH 107 (ATCC #31538 [bacteria with plasmid]; ATCC #40011 [plasmid only])

pGEM T-Easy (Promega #A3160) (comes with T4 DNA ligase and buffer)

pG1-21-4 (ATCC #63229)

pGEMEX-1 and pGEMEX-2 (Promega #P2211 and #P2551)

pYNH24 (D2S44) (ATCC #57570 [bacteria with plasmid]; ATCC #57571 [plasmid only])

pHY10 (DYZ1) (ATCC #57498 [bacteria with plasmid]; ATCC #57499 [plasmid only])

pARC145G (ATCC #40753)

pARC1520 (ATCC #40852)

pWE15 (Stratagene #251301)

Premixed or Premade Reagents

30% acrylamide-bisacrylamide, 37.5:1(Bio-Rad)

Note: Precast gels are also available.

8% acrylamide-bisacrylamide, 19:1 (Bio-Rad)

Bradford reagent (Bio-Rad)

BCIP-NBT developing solutions (Bio-Rad)

Silver Sequence DNA sequencing system (Promega #Q4160 [20 reactions] or #Q4130 [100 reactions])

10 mM deoxynucleoside triphosphates (dNTPs) (Boehringer Mannheim, Promega, and others)

Gel-drying apparatus (Promega #V7120 and others)

Genius DIG DNA labeling and detection kit (Boehringer Mannheim #1-363-514)

Gigapack III XL packaging extract (Stratagene #200202)

E.Z.N.A. gel extraction kit (Omega Biotech)

Puregene System (Gentra Systems #D5000A)

N,N,N′,N′-tetramethylethylenediamine (TEMED) (Bio-Rad)

Wizard gDNA Purification System (Promega #A1120)

Wizard DNA Clean-up System (Promega #A7280)

Oligonucleotides

TPA-25 Primer Sequences

5′ GTA AGA GTT CCG TAA CAG GAC AGC T 3′

5′ CCC CAC CCT AGG AGA ACT TCT CTT T 3′

APOC2 Primer Sequences

5′ CAT AGC GAG GAC TCC ATC TCC 3′

5′ GGG AGA GGG CAA AGA TCG AT 3′

D1S80 Primer Sequences

5′ GAA ACT GGC CTC CAA ACA CTG CCC GCC G 3′

5′ GTC TTG TTG GAG ATG CAC GTG CCC CTT GC 3′

Human Beta-Globin (BGP) Primer Sequences

BGP1: 5′ CAC CTG ACT CCT GA 3′

BGP2: 5′ AAT AGA CCA ATA GGC AGA G 3′

Small Ribosomal Subunit (SSU) Primer Sequences
SSU-1: 5′ GTG GAT CCA TTA GAT ACC C

SSU-2: 5′ ACT GGT ACC TTG TTA CGA CTT

Oligonucleotides are available from Life Technologies and many other companies.

- Oligonucleotides are generally received lyophilized (freeze-dried) from the supplier. Dilute them with 1× TE or dH_2O to a concentration of 100 pmol/µl. Store the concentrated stock at −20°C. Make a diluted stock of 10 pmol/µl, also stored at −20°C, for use when making PCR master mix. (Oligonucleotides are most stable when stored concentrated at −20°C.)
- Prepare PCR master mix in multiples of 25-µl aliquots at 2× final concentration. Students will add 25 µl total of DNA or DNA + dH_2O so that all reagents will be the final desired concentration.
- A general formula for PCR master mix is shown below.

Reagent	Stock	Amount per 25-µl PCR master mix aliquot	Final
PCR buffer	10×	5 µl	1×
BSA	1 mg/ml	0.5 µl	0.5 µg
dNTP mix	10 mM each	1 µl	200 µM
Primer #1	10 pmol/µl	2.5 µl	25 pmol
Primer #2	10 pmol/µl	2.5 µl	25 pmol
Taq polymerase	5 U/µl	0.5 µl	2.5 U
dH_2O		13 µl	
DNA		0 µl	25 µl
Total		25 µl	50 µl

Biotechnology and Laboratory Suppliers

Accurate Chemical and Scientific
Phone: 1-800-645-6264
300 Shames Drive
Westbury, NY 11490
Antibodies

American Type Culture Collection (ATCC)
Phone: 1-800-638-6597
10801 University Blvd.
Manassas, VA 20110-2209
Repository of various cultures, including bacteria, cell lines, DNA clones, and many others

Barnstead/Thermolyne
Phone: 1-800-553-0039
Fax: 1-610-250-1291
2555 Kerper Blvd.
P.O. Box 797
Dubuque, IA 52004-0797
Biotechnology equipment

Bio101
Phone: 1-800-424-6101
P.O. Box 2284
La Jolla, CA 92038
Reagents

Bio-Rad Laboratories
Phone: 1-800-424-6723
Phone: 1-510-741-1000
2000 Alfred Nobel Drive
Hercules, CA 94547
Enzymes, biotechnology reagents, and equipment

Boehringer Mannheim
Phone: 1-800-262-1640
Fax: 1-800-428-2883
P.O. Box 50414
Indianapolis, IN 46250-0414
Enzymes and chemicals

Brinkmann Instruments, Inc.
Phone: 1-800-645-3050
One Cantiague Road
P.O. Box 1019
Westbury, NY 11590-0207
Equipment and supplies

Carolina Biologicals
Phone: 1-800-334-5551
2700 York Road
Burlington, NC 27215-3398
Educational company

Clontech
1020 East Meadow Circle
Palo Alto, CA 94303
Enzymes and reagents

Difco Laboratories
Phone: 1-313-961-0800
P.O. Box 1068
Detroit, MI 48232
Microbial media

Fisher Scientific
Phone: 1-800-766-7000
52 Fadem Road
Springfield, NJ 07081
Supplies, reagents, enzymes, and equipment

Note: Their catalog has a specific biotechnology section that helps narrow purchasing choices.

Foto-Dyne
Phone: 1-800-362-3686
950 Walnut Ridge Drive
Hartland, WI 53029
Equipment

Gentra Systems
Phone: 1-800-866-3039
P.O. Box 13159
Research Triangle Park, NC 27709
DNA purification reagents

Kauai Fruit and Flower Company
Phone: 1-800-229-1814
Fax: 808-245-5325
E-mail: pineapple@hawaiian.net
3-4684 Kuhio Highway
Lihue, Kauai, HI 96766
Papaya fruit

Life Technologies, Inc. (Gibco/BRL)
Phone: 1-800-828-6686
P.O. Box 5009
Gaithersburg, MD 20877
Media, reagents, and oligonucleotides

MBI Fermentas
Phone: 1-800-340-9026
Suite 29
4240 Ridge Lea Road
Amherst, NY 14260
Enzymes and reagents

Millipore Corporation
Phone: 1-800-645-5476
80 Ashby Road
Bedford, MA 01730
Water purification equipment

New England BioLabs
Phone: 1-800-632-5227
32 Tozer Road
Beverly, MA 01915-5599
Enzymes and reagents

Omega Biotek
Phone: 1-888-296-6342
Suite G
6050 McDonough Drive
Norcross, GA 30093
Biotechnology reagents

Promega Corporation
Phone: 1-800-356-9526
2800 Woods Hollow Road
Madison, WI 53711-5399
Enzymes and reagents

Qiagen
Phone: 1-800-426-8157
9600 DeSoto Road
Chatsworth, CA 91311
DNA chromatography

Rainin
Phone: 1-800-828-2788
Fax: 1-617-938-8157
Mack Road
P.O. Box 4626
Woburn, MA 01888
Micropipettes and tips

Sigma
Phone: 1-800-325-3010
P.O. Box 14508
St. Louis, MO 63178
Biochemicals and reagents

Stratagene
Phone: 1-800-424-5444
11011 North Torrey Pines Road
La Jolla, CA 92037
Enzymes, reagents, and equipment

Worthington Biochemical Corporation
Phone: 1-800-445-9603
Fax: 1-800-368-3108
730 Vassar Ave.
Lakewood, NJ 08701
Chemicals and antibodies